Praise for *Hope Matters*

"Like Elin, I have met countless people who have lost hope in many countries. Most were apathetic. Some were angry. Others depressed. Because, they told me, their future has been compromised and there was nothing they could do about it. But there is something they can do. This book comes at just the right moment. It brings a message of hope to help curb the negativity, the gloom and doom we are confronted with each day. It is NOT too late if we get together and take action, NOW."

JANE GOODALL, PhD, DBE, founder of the Jane Goodall Institute and UN Messenger of Peace

"This is a truly eye-opening book: about our endangered planet, with many signposts towards a better world. Kelsey's study is full of illuminating analyses and uplifting, empowering stories about people from all over the globe. A beautifully written book and an effective antidote against apathy and inaction."

CHRISTOF MAUCH, director, Rachel Carson Center for Environment and Society

"At a time of overwhelming anguish for our fellow citizens and the health of our planet, Kelsey provides clear strategies to translate hope into action, with inspiring stories of ecological resilience and restoration across the globe. *Hope Matters* points the way forward."

FRANCES BEINECKE, former president, Natural Resources Defense Council

"Elin Kelsey writes with the acuity of a scientist, the grace of a poet, and the heart of a mother. *Hope Matters* is a clarion call to reawaken our spirits and renew our efforts—a book to inspire resilience for our children, for our leaders, and for ourselves."

ANNE NELSON, Fellow, Arnold A. Saltzman Institute, Columbia University School for International and Public Affairs

"In a time when so much of the news on biodiversity is depressing, Kelsey reminds us that there are good reasons to be hopeful. This book is a tonic in hard times."

CLAUDIA DREIFUS, author of *Scientific Conversations: Interviews on Science from the* New York Times

"As we work to turn back the urgent threats of climate change and species extinction, Kelsey shows through countless well-chosen examples that the solutions we need are there if we learn the habits of optimism and practicality rather than succumbing to despair."

GEORGE BLACK, author of *Empire of Shadows: The Epic Story of Yellowstone*

"Required reading for anyone despairing about the future of the planet."

MITCHELL THOMASHOW, author of *To Know the World: A New Vision for Environmental Learning*

"This book has an important message for everyone, including (and perhaps especially) for the hope skeptics. Hope is not about feeling cheerful or ignoring the facts to look on the bright side; rather, it's about 'recognizing both the threat and the potential solvability of the climate crisis' and amplifying transformative solutions."

VERONICA JOYCE LIN, North American Association for Environmental Education "30 Under 30" education influencers

"A refreshing and uplifting read. It is so useful as a teacher to have information to share with students about the power of the collective, the power of the small gesture which eventually is magnified, and the beauty, power, and creativity of the living, breathing world. An amazing, powerful, and wonderfully heartwarming book!"

ENID ELLIOT, leader in the Nature Kindergarten movement

"An uplifting and eye-opening read for anyone worried about climate change and the state of our planet. Kelsey brings us stories of recovery and resilience. A must-read for anyone writing or reading about the perils of climate change."

LESLIE KRAMER, former producer, *CBC News*

HOPE
MATTERS

Why Changing the Way We
Think Is Critical to Solving
the Environmental Crisis

ELIN KELSEY

DAVID SUZUKI INSTITUTE

GREYSTONE BOOKS
Vancouver/Berkeley

20 21 22 23 24   5 4 3 2 1

Greystone Books Ltd.
greystonebooks.com

David Suzuki Institute
davidsuzukiinstitute.org

Cataloguing data available from Library and Archives Canada
ISBN 978-1-77164-777-9 (pbk)
ISBN 978-1-77164-778-6 (epub)

Editing by Paula Ayer
Copy editing by Janine Young
Proofreading by Jennifer Stewart
Indexing by Susan Safyan
Cover design by Jennifer Lum
Text design by Nayeli Jimenez
Cover photograph by iStockphoto.com
Printed and bound in Canada on ancient-forest-friendly paper by Friesens

Greystone Books gratefully acknowledges the Musqueam, Squamish, and Tsleil-Waututh peoples on whose land our office is located.

Greystone Books thanks the Canada Council for the Arts, the British Columbia Arts Council, the Province of British Columbia through the Book Publishing Tax Credit, and the Government of Canada for supporting our publishing activities.

BRITISH COLUMBIA

BRITISH COLUMBIA ARTS COUNCIL
An agency of the Province of British Columbia

Canada Council for the Arts   Conseil des arts du Canada

A Note About Time

This book will not go out of date because the tension between hope and despair is always with us. At the same time, all of the examples I use are real life issues that are rapidly changing. I time-stamped content throughout the chapters to give you a sense of the current state of play when the book was written. Please be sure to hop online and look for positive developments that have transpired since I wrote this.

But, there are bats
who catch fish
and slime molds that sing

and ancient Greenland sharks
who don't even reach sexual maturity
until they are 150 years old.

This is what I want to say
to those who believe
the Earth is already doomed:

Just look
at the capacity
of this
gloriously complex planet.

CONTENTS

INTRODUCTION

I WROTE THIS BOOK for you—and for the people you love who believe the world is screwed.

I suspect you know who I mean—the ones brave enough to acknowledge the existential angst of living and working through a planetary emergency. We see it in students who appear to have everything going for them yet are so desensitized by environmental despair, they are simply done with feeling; or the ones so saturated with doom-and-gloom messages, they honestly believe no future exists. We hear it in adults grieving for the world their children will live in. We watch it echo around the world through the voices of the million and a half students joining Greta Thunberg in school strikes against climate change.

Whenever I speak about environmental solutions and positive global conservation trends, I encounter people so hungry for hope, they line up to talk to me. They follow up with long, thoughtful emails telling me they honestly believe the state of the planet is past the point of no return. They express incredulity that hope could exist.

What is clear from these conversations is that more of us are aware of the very real and urgent environmental problems we face than ever before; more people care and are ready to take action, and our worries about these issues are profoundly impacting our lives. According to the American Psychological Association, climate change has far-reaching effects on our mental health: it triggers stress, depression, and PTSD, strains our personal and community relationships, and leads to increases in aggression, crime, and violence, to name but a few of these effects.

My journey into this issue began in earnest in 2008, when I led a series of workshops at the United Nations Environment Programme's Children's Conference on the Environment in Stavanger, Norway. I met with children between the ages of ten and fourteen, from ninety-two different countries. When I asked them, *What words or feelings come to mind when you think about the environment?* I was overwhelmed by their expressions of fear, anger, and despair.

Kids take their cues about the state of the planet from adults and the media. Ironically, the ways in which we talk about the environment—chronicling its demise, making threatening forecasts about the future, blaming others, framing things in terms of "never" and "always," and telling others what they *should* do—parallel the dysfunctional ways we so often communicate in our most intimate relationships. Curating the bad stuff into a display of hopeless futures is just as useless when it comes to engaging people with the planet as it is with affairs of the heart. Badgering someone you love to change will not fix things. Nor can we save the world by continuing to focus on what people are doing wrong.

Yet the narrative of gloom and doom continues to dominate how we think, feel, and relate to the environment the world over. That's because so many people fear that focusing on solutions could unintentionally fuel complacency or give politicians an out, or that we might champion a solution that isn't letter-perfect. So we continue to reproduce this culture of hopelessness in spite of the mounting evidence of how damaging it is to our personal health and well-being—and to our collective capacity to respond to urgent environmental issues.

I often wonder why we treat the planetary emergency so differently than we do other terrifying events. Time and again, we see communities come together in response to a school shooting or mass flood. We see adults reassuring children that steps are being taken to make them safe or avoiding exposing young children to news items beyond their grasp. Even in the ordinary circumstances of everyday life, we have age restrictions on violence in movies. Yet when it comes to the environment, we bombard kids with horrifying content about the ruined state of the planet their lives depend on, without any support for how those messages make them feel. Within the context of environmental issues, we seem to flip the story: we shrug off our moral responsibilities as adults to help younger people through their suffering; instead we tell children that it is up to them to save the world.

Somehow, we've got urgency, crisis, and fear all balled up. Environmental issues *are* real, and they *are* at the level of global crises. But failing to separate the urgency of these problems from the fear-inducing ways in which we communicate them blinds us to the collateral damage of apocalyptic storytelling. We grow deaf to more inspiring and effective possibilities. By hammering children and adults with issues at scales that feel too large to surmount, we inadvertently cause them to downplay, tune out, or shut down. We are fueling an epidemic of hopelessness that threatens to seal the planet's fate. The environmental crisis is also a crisis of hope.

THIS BOOK SEEKS to right that wrong. Within these pages, I craft an evidence-based argument for hope that reflects the complex psychological, sociological, philosophical, and spiritual qualities of this phenomenon. I share insights about hope and the environment that have been honed through thousands of conversations with people around the world who have generously trusted me with their feelings. It's an "eyes-wide-open" look at hope within the full recognition of the gravity, urgency, and vast entanglements of the planetary crisis.

The situation of hope vis-à-vis the environment is particularly tricky. Whereas a patient in hospital may experience hope within a realistic understanding of even limited treatment options, those suffering from hopelessness about the environment often have little or no notion that environmental successes exist. Because they *feel* hopeless, they believe the situation *is* hopeless.

In these pages, I argue that hope for the environment is not only warranted but essential to addressing climate change, biodiversity loss, and the full suite of environmental crises we face. By focusing our attention so heavily on what's broken, we are reinforcing a starting-line fallacy that makes it feel as if nothing useful has ever been accomplished and that all the hard work lies ahead. We need to pry ourselves free from this disempowering rhetoric and situate ourselves within the positive environmental trends that are already well established and yielding the successful results we need to grow.

Turning toward solutions is a tall order. For starters, environmental solutions aren't easy to find. Environmental news is almost exclusively reported as "bad" news. When we do come across a positive environmental story in the media, it's frequently presented as a one-off good-news story. Too often, this positions environmental solutions as rare exceptions rather than examples of major trends that have been building over time.

Furthermore, feeling empowered to act demands a sense of possibility that is being constantly eroded by the now-ubiquitous exposure to horrifying events happening around the world daily. The strains on human emotions are far greater than ever before thanks to social media, twenty-four-hour news cycles, and alerts on personal devices. Relentless exposure to widespread tragedy fuels emotional exhaustion, leading to desensitization, cynicism, and resistance to help those who are suffering.

FOR YEARS, I would ask every scientist I met what made them feel hopeful about the environment. They always had an answer. I mistakenly thought that the only way to increase exposure to under-reported successes would be for me to interview these folks and gather examples one by one. Luckily, in April 2014, I met Patrick Meier, a pioneer in the rapidly developing world of digital humanitarianism, while we were fellows together at the Rockefeller Foundation's Bellagio Center in Italy. When he described how he mobilized the Haitian diaspora after the 2010 earthquake to create and interpret a crisis map for rescue workers from the tweets, text messages, emails, and Facebook posts coming from people on the ground, I began to imagine ways we might use social media to gather specific cases of environmental successes from all over the planet and to make them easily shareable. We could try to crowdsource hope.

The result was #OceanOptimism, a social media campaign designed to crowdsource and share examples of ocean conservation successes and solutions. We launched it on World Oceans Day in June of 2014, and to the surprise of our tiny group behind the campaign, the tag went viral, reaching more than ninety million shares to date. Within seconds of logging onto Twitter or Instagram, users see current accounts of ocean successes: whales returning to New York City harbor, or the resilience of corals on a damaged deep-sea reef off Scotland. They find inspiring surprises, like the discovery of 200,000 cold-water coral reefs off the coast of Norway, or the recovery of groundfish along the California coast—some populations rebounding fifty *years* sooner than predicted.[1]

Knowing what works matters. These social media feeds make searching for replicable solutions or finding people working on similar issues much easier, which translates into successful actions being reproduced and tailored to other situations. And daily exposure to posts of conservation successes changes how people feel about the state of the planet. These successes empower us. Innovations and positive feelings, in turn, spread.

The mass accumulation of all these examples now represents valuable data sets, which enable researchers to look not only at specific examples but also at broader trends. In chapter eight I'll point out some promising examples, inviting you to see how the issues you care about exist within much larger movements of positive change.

Social and technological revolutions surround us. Remote sensing, big data, and a suite of technologies provide new ways to understand the 8.7 million other species on Earth. Animal-tracking devices are becoming ever lighter, sturdier, smaller, cheaper, and better able to store and transmit large amounts of data. Scientists can now affix tiny solar-powered tracking devices onto the slim legs of songbirds, for instance, and use their personal devices to track how certain species "surf the green wave," timing their migrations to the arrival of spring across continents.[2] Precision conservation technologies pinpoint areas of highest biodiversity importance, charting areas of greatest vulnerability and providing evidence to support conservation protection.

We are living in an Age of Personalization where, thanks to eBird, Song Sleuth, iNaturalist, or other environmental apps on our personal devices, we can now identify those same birds with the touch of a screen. By holding our iPhones up to our binoculars, we can easily take a close-up photo, revel in the beauty of a bird's colorful feathers, and post the image on Instagram. This enhanced capacity to *see* birds is fueling an international urban birding movement, especially among folks in their teens, twenties, and thirties. And that turns out to be a boon for conservation. Birders, according to a recent study, are five times more likely than non-birdwatchers to participate in wildlife and habitat conservation; to recycle and engage in other eco-friendly activities; and to vote for wildlife-positive regulations and policies.[3]

The tired old narrative of doom and gloom can no longer capture the changing global dynamics of life on planet Earth. The constant harkening back to fear does not serve us.

Far from keeping us from growing complacent, fear drives apathy. Indeed, there is growing concern that framing climate change as an impending catastrophe stokes the fires of "climate doomism"—the fatalistic belief that it is already too late to act, which, according to researchers, causes people to give up. And conversely, when we act from a positive feeling of meaning and purpose, we all benefit. In 2013, in the first study of its kind, medical researchers demonstrated that the happiness we derive when we act on behalf of the greater good shows up in our cells as a healthier immune response.[4]

Hope Matters is a timely, evidence-based argument for the place of hope and a celebration of the turn toward solutions that is emerging in the face of global crisis and despair. I hope you will share it with the people you love.

1

THE POWER OF
EXPECTATION AND BELIEF

What we pay attention to
shapes our lives
no matter what species we are.

Plant roots
sense and sidestep rocks
before they hit them.

Songbirds
avoid tornados
by listening for them
from hundreds of miles away.

And elephants
sneak their way into fields,
by watching—then imitating—
how the farmers
cross the moats and fences
they've built
to protect the crops.

W E ARE LIVING amid a planetary crisis. "I am hopeless," a student in an environmental study graduate program recently told me. "I've seen the science. I am hopeless because the state of the planet *is* hopeless."

It's not surprising she feels so depressingly fatalistic. In his speech at the start of a two-week international conference in Madrid in December 2019, UN Secretary-General Antonio Guterres said, "The point of no return is no longer over the horizon. It is in sight and hurtling toward us."[1] And this student isn't alone in her feelings. I often give public talks and no matter where I am in the world, I begin by inviting people to share how they are feeling about the environment with the person sitting beside them, and then, if they are willing, to call out some of the words that capture these feelings. I have done this hundreds of times, and every time, the answers shock me. When I look out at these audiences, I see bright, healthy, relaxed-looking people who have somehow found the time to come to a public lecture. Yet their answers convey an unnerving level of grief and despair: "Scared," "Hopeless," "Depressed," "Numb," "Apathetic," "Overwhelmed," "Guilty," "Paralyzed," "Helpless," "Angry," call out the voices. Whether the room is filled with adults, university students, or kids as young as grade three, whenever I ask, the words remain the same.

Not long ago, I found an almost identical collection of words. It's a list published in a research journal by Johana Kotišová. The words describe the emotions that crisis reporters feel when they are covering horrific events such as the Haitian earthquake, the Brussels or Paris attacks, the war in Ukraine, the war in Liberia, refugee camps, 9/11, famines in Central African countries, or the aftermath of the Greek debt crisis.[2]

The same words. What I am saying is that ordinary kids and adults regularly describe their everyday feelings about the environment using the same words that journalists use to describe what it feels like to report on the worst imaginable crises.

The environmental crisis is also a crisis of hope.

This crucial idea drives this book. My agenda is absolutely to spread hope. I believe the way to do that is to collectively challenge the tired narrative of environmental doom and gloom that reproduces a hopeless status quo, and replace it with an evidence-based argument for hope that improves our capacity to engage with the real and overwhelming issues we face.

The power of hope and beliefs

When everyone around you is shouting doom and gloom, actively choosing to be hopeful—and to do the hard work of seeking out and amplifying solutions—is difficult. But it's also essential, because hope really matters.

It matters to your individual health and well-being. Many studies underscore the value of feeling hopeful in all sorts of situations. If you have hope, you're better able to tolerate pain. You're more likely to follow through with physiotherapy or other recuperative treatments following an injury or illness. Feeling hopeful leads to better recovery from anxiety disorders and cardiovascular disease.[3] The capacity to hope has been shown to provide a therapeutic quality that helps refugees overcome seemingly insurmountable challenges as they move forward and resettle.[4]

Being hopeful also matters to how we collectively influence what happens on the planet. That's because thoughts, feelings, perceptions, and beliefs are so powerful, they actually shape objective outcomes.

The placebo effect

I had the opportunity to think more about the power of expectations when I was a visiting scholar at Stanford University in 2018. A researcher named Parker Goyer at the Mind & Body Lab generously talked me through the breakthrough work the lab's founder, Alia Crum, and her team were doing.

You're probably familiar with the placebo effect. Though not named as such at the time, the concept dates back almost five hundred years. "There are men on whom the mere sight of medicine is operative," wrote the French philosopher Michel de Montaigne in 1572,[5] referring to situations in which a person experiences relief from pain, anxiety, or other symptoms because they believe they have taken medicine or received treatment, when in reality, they have not.

The placebo effect demonstrates the power of our minds to produce physical changes in our bodies. Clinical trials demonstrate that if we believe we are taking a real medication, then something as simple as taking a sugar pill can lower our blood pressure, reduce anxiety and pain, and boost our immune system.[6]

Placebos work by triggering a host of specific neurobiological effects. As Alia Crum explains it, "The power is not in the sugar. The power comes from the social contexts that shape our mindsets in ways that activate our bodies' natural healing abilities."[7]

Think of a mindset as a lens or frame through which we view the world. Our mindsets orient us to particular associations and expectations. Our mindsets don't just color our reality. Rather, the way that we look at reality changes what we pay attention to, and what we expect. Believe it or not, those expectations and associations actually change that reality.

Mindsets impact objective reality

Take this study involving hotel cleaning staff. Early in her academic career, Alia worked with Harvard University psychologist Ellen Langer on a study that involved women hotel-room attendants. Cleaning all day involves lots of physical activity, but the women doing that job didn't think of their work as good exercise. The researchers divided the hotel-room cleaners into two groups. In one group, they did no intervention. But with the second group, they showed the women how the work they did cleaning actually more than met the US surgeon general's recommendation for daily physical exercise: detailing, for example, how fifteen minutes of vacuuming burns fifty calories, fifteen minutes of scrubbing sinks burns sixty calories, and so on. They posted this information in the staff areas at the hotels where only those room attendants in the second group would see it.[8]

A month later, the researchers checked back. That simple intervention—no changing of diet or exercise regime, just promoting the mindset that "work is good exercise"—produced dramatic results. Hotel-room attendants in the second group lost weight and lowered their blood pressure on average by ten points.[9]

These findings demonstrate the capacity of our inner dialogues and self-perceptions to manifest themselves. Objective health benefits depend not just on what we do, but what we *think* about what we do.

These days, many of us know the benefits of a plant-based diet for our health and the environment. So, what's the best way to help someone choose to eat vegetables? The most common method for encouraging healthier food choices is to prominently display nutrition information. But Alia and other researchers at the Mind & Body Lab found that focusing on health but failing to mention taste unintentionally instills the mindset that healthy eating is flavorless and depriving.

In 2016, they tried a new approach and applied it to the food sold on the Stanford campus. The researchers chose adjectives that popular restaurants used to describe tasty but less healthy foods, and then used those same words to name vegetable dishes that were both nutritious and tasty.

Decadent-sounding labels—like "twisted, citrus-glazed carrots" and "ultimate chargrilled asparagus"—persuaded more people to choose veggies.

They took the study nationwide, testing the same idea in fifty-seven US colleges and universities. They tracked nearly 140,000 decisions about seventy-one vegetable dishes. It turns out diners put vegetables on their plates 29 percent more often when those vegetables had tasty-sounding labels than they did when the vegetables had health-focused names, and 14 percent more often than when the veggies were given neutral names.[10]

Yummy labelling works because it makes eating healthy crave-worthy. Knowing that veggies are healthy and that eating them is the right thing to do isn't enough. We are more likely to do something good when it also feels good. Our feelings about eating vegetables are not fixed. It's not that we either love or hate vegetables; rather, our decision to eat them is influenced by labels that appeal to a delicious and indulgent mindset.[11]

People, too, can exert a placebo effect. When British doctors in a now-famous empirical study gave patients (who were suffering from minor cold symptoms or mild muscle pain) a firm diagnosis and positive assurances that they'd feel better in a few days, 64 percent of those patients got better. But when patients with the same symptoms were seen by doctors who told them they were uncertain of the diagnosis, and that if the patient still felt ill in a few days they should return to the doctor, only 39 percent said their health had improved.[12]

What we expect can cause negative consequences. Back in 1962, Japanese researchers did an experiment on thirteen boys who were hypersensitive to the leaves of the Japanese lacquer tree.[13] Contact with leaves from these trees can cause a painful, itchy rash similar to poison ivy. The researchers touched the boys on one arm with leaves from a harmless tree and told them they were from the Japanese lacquer tree. They touched them on the other arm with leaves from the Japanese lacquer tree but told them the leaves were harmless.

All thirteen arms that had been touched by the harmless leaves showed a skin reaction. Only two of the arms that were touched by the poisonous tree produced a rash. Even more surprising, the reaction to the *harmless* leaves was stronger than the reaction to the leaves that were actually poisonous. Simply thinking that one is being touched by a poisonous leaf brought on a rash more often than actually being touched by one. Health professionals sometimes see the same phenomenon happen with patients who fear uncomfortable side effects to a prescription. The capacity of inert substances to bring about pain and other negative responses, simply because we expect them to do so, is called the nocebo effect.[14]

So, what does all of this have to do with hope and the environment?

Suffering headline stress disorder

Think of the environmental stories you've consumed recently. How do they make you feel? What's the impact of being bombarded by the climate crisis, species extinctions, wildfires, plastics pollution, and so many other urgent, global issues?

We are exposed to horrifying events more today than at any other time in human history. Twenty-four-hour news cycles, alerts on personal mobile devices, and social media feeds bring incessant predictions of a bleak future.[15] The percentage of adults using social networking sites jumped tenfold in the past decade. Much of our news consumption now occurs on these digital platforms. A mobile phone image taken by Alexander Chadwick, a survivor of the 2005 London subway bombings, jump-started what is now the everyday practice of reporting news in part through user-generated content. Our increased exposure to real-time, on-the-ground knowledge of things happening all over the planet can help build connections with people in different circumstances, but it also places us in perpetual, intimate contact with tragedies, which leaves many feeling cynical, desensitized, and ineffectual. Life has always been stressful and terrible things certainly happen, but personal exposure to horrifying events occurring any place on Earth is a new and disturbing phenomenon. What is also new is our heightened exposure to images and videos captured by ordinary people on their smartphones detailing the devastation of climate change, along with a clear message from our most trusted scientific sources that if we do not act fast, even more dire consequences are coming.

The anxiety, exhaustion, and difficulty sleeping many experience in response to the news has become so prevalent in recent years, psychologist Steven Stosny gave it a name: *headline stress disorder*. It's the state of anxiety and fear people experience in response to an intense deluge of terrible news. Caught in a self-perpetuating cycle of doom and gloom, people experience a range of emotions, including fear, anxiety, anger, and depression. As Steven Pinker puts it, "Whether or not the world really is getting worse, the nature of news will make us think it is."[16]

It's a matter of quantity—seven in ten Americans say they feel worn out from too much news, according to a 2019 Pew Research Center study—and orientation. As numerous communication studies reveal, almost all of the news that we hear about the environment is bad. It feels like the world is falling apart.[17]

That's a problem for you, and for everyone in your social network. Emotions are contagious. Not only in face-to-face situations, but online too. Every time you click on a terrifying news story about the state of the planet on social media, you are actively "catching" emotional despair, and every time you post or share that message, you're spreading it.

We hear a lot more news about environmental problems than solutions

Climate change is an urgent, global-scaled problem. In October 2018, the world's leading climate scientists—the Intergovernmental Panel on Climate Change—released their most dire report ever: the world is currently 1 degree Celsius (1.8°F) warmer than preindustrial levels, and every fraction of additional warming will worsen the devastating impact of climate instability.

Yet worrying about a problem that is *way* too big for you to tackle inevitably feels discouraging. It's disempowering. It breeds apathy. The same phenomenon happens in politics.

When someone says, "Why would I bother voting?" they may be finding it hard to see how their single ballot among thousands or millions makes a difference.

To counter this feeling, psychologists say it's important to see how our individual actions make a collective positive impact.[18] Indeed, research demonstrates that when the news focuses on success stories about entrepreneurial activism and actions ordinary people are taking in local contexts we can relate to, we feel more enthusiastic and optimistic about our capacity to tackle climate change.

But unfortunately, that's not the way climate change is typically reported. Less than 19 percent of climate change coverage on major nightly news programs in the US in 2017 and 2018 mentioned climate change solutions.[19] We might assume that negative news will shock people into action, but instead it's been proven that it can cause them to disengage. Stories that emphasize the failures of climate politics intensify people's feelings of despair and cynicism. Journalist Elizabeth Arnold, in her five-year study of national media coverage about climate change in the US Arctic, found that almost every story perpetuated a narrative of "fear, misery, and doom" that left the public feeling powerless.[20] The effects of this on our personal health and well-being are profound. As David Bornstein (journalist and co-creator of the Solutions Journalism Network) put it: "If the news were a pill, and all the known effects of the news were given in pill form, the FDA probably wouldn't approve it."[21]

In a 2017 review of more than fifty thousand abstracts from articles published in ocean and coastal science journals between 2006 and 2015, Murray A. Rudd of Emory

University determined that the vast majority of articles did not propose actual solutions to environmental-change challenges. Because environmental reporters often base their reports on journal findings, those reports are heavily weighted toward presenting problems without solutions.[22]

In many ways, the negative skew of climate change media stories is also a reflection of the general tendency for the media to focus on negative news. Plane crashes, for example, are always covered in the news, but car crashes hardly ever are, even though they kill more than 125 million people (and injure and maim 20–50 million more) every year.[23] The likelihood of dying in a plane crash is extremely low. In 2019, the fatal accident rate was on average one death for every 5.58 million flights.[24]

Studies reveal that news all over the world has grown gloomier in the past two decades. Major US newspapers, studies show, are far more likely to report on unsuccessful climate actions than they are to cover climate action successes.[25] The same is true internationally. Maxwell Boykoff directs the Center for Science and Technology Policy Research at the University of Colorado Boulder. The Center operates a Media and Climate Change Observatory, which monitors how climate change is reported across 117 sources (newspapers, radio, and TV) in fifty-five countries. They've found that problems caused by climate change are deemed more newsworthy than solutions, and that this coverage drives a sense of hopelessness. "There's still a pervasive doom and gloom," Boykoff said in a 2018 interview. "When these stories just focus in on doom and gloom, they turn off those who are consuming them. Without being able to find their own place as a reader, viewer, or listener in those stories, people feel paralyzed and they don't feel like they can engage and have an entry point into doing something about the problem."[26]

These findings are worth paying attention to because the number one way most of us learn about the environment is through the media. Media shape the stories we hear, which, in turn, become the mindsets that we use to understand the world.

Catastrophe narratives in pop culture

Climate change fatalism is so ubiquitous it's made its way into pop culture. In the HBO series *Euphoria*, for example, a teen addict defends her drug use, saying: "The world's coming to an end, and I haven't even graduated high school yet." It's just one of a seemingly endless stream of apocalyptic and post-apocalyptic films and television series that emerged over the past decade. The surge of catastrophe narratives led *New York Times* film critic Anthony "Tony" Scott to quip, "Planetary destruction and human extinction happen a half-dozen times every summer" in his 2014 review of the movie *Snowpiercer.*[27]

Popular culture provides a lens through which we can see how, as a broader society, we are thinking and feeling. It both influences and reflects societal concerns and desires. Fears about climate change, and the profound ecological uncertainty and change it engenders, are so resonant they've given rise to a whole new genre of ecological-disaster-themed entertainment, commonly referred to as "eco-apocalypse," "eco-catastrophe," or "climate porn."[28]

In 2019, Shauna Doll and Tarah Wright of the Education for Sustainability Research Group at Dalhousie University[29] did a thematic analysis of two hundred artworks related to climate change from across Europe and North America. Only four artworks were coded as expressing "hope." This is a problem, particularly given the findings of researchers from the Norwegian University of Science and Technology. They studied the impact of the art displayed in Paris in association with the 2015 United Nations climate change summit. They too found that the vast majority of the pieces were dystopian and gloomy, and that those works left people feeling uninspired to take action. Only three of the thirty-seven works on display left people feeling hopeful that they could do something about climate change—all three of those works focused on solutions.[30]

Any narrative that is so deeply embedded should raise alarm bells. There should never be just one dominant story. In a well-functioning, democratic world, there are multiple stories competing with one another for our attention. The idea that something as complex and extraordinary as all life on Earth could ever be encapsulated by a single grand narrative just doesn't make sense. It's as if "the Earth is dying" has become a sort of apocalyptic platitude. We repeat these things because we've heard each other say them, but it's possible we say them without really thinking about what they actually mean. We have massive, terrifying, urgent environmental problems. But we also have powerful successes that we need to amplify above the din of hopelessness.

Whenever we straitjacket an idea or an issue into a single, monolithic story, whether it's "environment" or "Africa" or "gay" or "terrorist," we lose the nuance and specificity of context. We miss positive developments and shifts in perception. We are left with an oversimplification that is so generalized it becomes inherently inaccurate. Because we are told that the planet is doomed, we do not register the growing array of scientific studies demonstrating the resilience of other species. For instance, climate-driven disturbances are affecting the world's coastal marine ecosystems more frequently and with greater intensity. This is a global problem that demands urgent action. Yet, as detailed in a 2017 paper in *BioScience,* there are also instances where marine ecosystems show remarkable resilience to acute climatic events. In a region in Western Australia, for instance, up to 90 percent of live coral was lost when ocean water temperatures rose, causing the corals to jettison the algae (zooxanthellae) living in their tissues—what scientists call coral bleaching. Yet in some sections of the reef surface, 44 percent of the corals recovered within twelve years. Similarly, kelp forests hammered by three years of intense El Niño water-temperature increases recovered within five years. By studying these "bright spots," situations where ecosystems persist even in the face of major climatic impacts, we can learn what management strategies help to buffer destructive forces and nurture resilience.[31] Rarely do the media return to profile the astonishing return of life after a catastrophic event.

Beware fatalistic mindsets

When the student I mentioned at the beginning of this chapter said, "I am hopeless because the state of the planet is hopeless," she believed that to be true, and I felt sad for her suffering. But I also saw her statement as an example of just how taken-for-granted and powerful the mindset of doom and gloom is. She described both her hopelessness and the hopeless state of the planet as non-negotiable, fixed, facts—as reality. She wasn't saying, "I feel hopeless." She was saying, "I *am* hopeless." Just as she wasn't saying, "I am worried that the state of the planet is hopeless," she was saying, "It *is* hopeless."

The vast scale, complexity, urgency, and destructive power of biodiversity loss, climate change, and countless other issues are real. Yet assuming a fatalistic perspective and positioning hopelessness as a foregone conclusion is not reality. It is a mindset, and it's a widespread and debilitating one. It not only undermines positive change, it squashes the belief that anything good could possibly happen. Record-high numbers of Americans worry about climate change, but only 5 percent of them believe that humans can and will success- fully reduce it, according to a 2017 study by researchers at Yale University and George Mason University.[32]

We need to decouple the enormity of the crises we face from the ongoing construction of hopelessness. Doom and gloom is so synonymous with the environment, we fail to recognize it as a frame, as a way of seeing things, as a mind- set. The mindsets we hold influence the outcomes that will result. Whether we are consciously aware of them or not, our mindsets affect what we pay attention to. Mindsets change what we are motivated to do and even what we believe is possible. We need to remind ourselves of this over and over and over again because, as we'll see in the next chapter, our blindness to hope is extracting far too heavy a toll.

2

THE COLLATERAL
DAMAGE OF DOOM AND GLOOM

Everything is shifting.
Gone,
the familiar pattern
of ordinary.

I am grieving the loss
of my everyday normal
in the midst of a global redistribution
of the entire world's species.

A mass unraveling of relationships,
all of us,
out of sync.

THE BELIEF THAT fear is a better motivator than hope is amazingly pervasive when it comes to the environment. The funny thing is that it runs counter to our own experiences in other parts of our lives. Think back to how you've felt when you had the misfortune of working for a tyrannical boss or a professor hell-bent on using cutthroat exams to reduce the class size. What you no doubt experienced is that fear can be a great mechanism to alert you to situations where failure is unacceptable. But fear of failure doesn't propel you to greatness. In fact, fear leads most of us to panic. We can't think straight. We stop looking for creative solutions or imaginative ways forward. Students and employees make more errors when they are operating in cultures of fear; because everyone is afraid of screwing up and being found out, we hide our mistakes, which means no one can learn from the mistakes of others.

Trying to avoid failure is a familiar but ineffective strategy. Failure, it turns out, is an essential prerequisite for success, according to a massive study of three-quarters of a million grant applications to the National Institutes of Health published in *Nature* in 2019. Yian Yin and his colleagues at Northwestern University set out to create a mathematical model that could reliably predict the success or failure of an undertaking. In addition to the grant applications, they also tested their model on forty-six years' worth of venture capital startup investments. The result?

Every winner begins as a loser. But the old proverb *if at first you don't succeed, try, try, again* only works if you learn from your previous failures. You need to keep doing what works and focus on changing what didn't. Plus, you should get right back up and try again. The more time you leave between attempts, the more likely you are to fail again. Rather than trying to avoid failure, what matters is what we learn when we fail, the changes we make based on that learning, and how quickly we try again.[1]

Fear can manifest as anxiety and hopelessness, which keeps us from being productive. Hopeful action, on the other hand, breeds confidence, happiness, and freedom to experiment—emotions that are tied to better performance and a better sense of well-being.[2]

Fear alone is not an effective strategy

Despite the well-documented ill effects of creating cultures of fear, I often meet people who believe fearmongering is necessary to spur environmental action. In fact, they tell me that the real problem is people aren't scared enough. Hope, they say, creates complacency at the very time we most need people to be scared into action. Clearly, that's the sentiment David Wallace-Wells channeled in his 2017 essay "The Uninhabitable Earth." The article delineates the effects of the worst-case scenarios of climate change, crafting a horrifying, dystopian vision of a near future destroyed by runaway climate change.[3] Within a week, it had become the most widely read article ever published in *New York* magazine.

There's no doubt fear makes a deep impression.[4] And it's true that fear-based messages can be effective, especially for simple, short-term, or specific behavior-changing interventions. Yet a 2014 meta-analysis that looked at the effectiveness of fear campaigns across sixty years of studies concluded that increasing people's confidence is a more successful approach than just trying to scare folks straight.[5]

Fear alone doesn't help us to address broad, complex, emotion-laden, societal-level issues, like the ones we face with climate change. Indeed, Anthony Leiserowitz, director of the Yale Program on Climate Change Communication, describes what he calls a "hope gap" between people's fear about climate change, and their feelings of powerlessness to do anything about it. Even those people identified as "most concerned about climate change" in research studies don't really know what they can do individually or collectively, he says. It's a serious problem. As Leiserowitz puts it, "Perceived threat without efficacy of response is usually a recipe for disengagement or fatalism."[6]

The hope paradox

We find ourselves, therefore, in a paradox. As I described in chapter one, climate change communication to date has overwhelmingly relied on negative emotions. One could argue it's been a highly successful tactic. American concern about climate change is higher than ever before, jumping 9 percent between 2018 and 2019.[7] Evidence from polls in many parts of the world indicates that concern about climate change is at a record high. Increasing numbers of people believe climate change poses a severe risk to themselves and the countries where they live, according to a survey of twenty-six nations conducted by the Pew Research Center in the spring of 2018.[8] Though the levels of concern vary by country, people rank climate change as the top global threat.

What all these polls confirm is that a critical mass of people all over the planet now know about and are also worried about climate change. This is an astonishing accomplishment. The effort required to focus global attention on a single issue is beyond challenging, especially for a problem as complex and difficult to communicate as climate change.

This mass demonstration of collective worry is driving political will toward change. Half of the world's population is younger than thirty years old. According to a 2019 World Economic Forum survey of thirty thousand individuals under the age of thirty across 186 countries, climate change and the destruction of nature is the biggest global concern for young people around the world.[9] The same year, a survey of over ten thousand eighteen-to-twenty-five-year-olds across twenty-two countries by Amnesty International found Generation Z fears climate change more than any other issue.[10]

We're beginning to see the results. For the first time ever, in 2019, climate change was a top issue in Canada's federal election. As I write this, the world's youngest prime minister, Sanna Marin, has set an ambitious target to make Finland the first carbon-neutral welfare state in the world. Bhutan and

Suriname have already won the net-zero-greenhouse-gas-emissions race, with Norway and Sweden coming up close behind. Meanwhile, in Australia, climate politics is burning as hot as the devastating bushfires.

Fear, guilt, and shame are powerful levers in political movements. However, fear tactics are a double-edged sword. On an individual level, fear is a good indicator that something is broken or has gone wrong. But, when it becomes entrenched, as it has in the doom-and-gloom narrative, it is demotivating. When we are afraid, we become less creative, less collaborative, and less capable of perseverance. And that's where the paradox comes in. As a global community, with climate concern at a record high, we are better positioned than ever before to take urgently needed action, yet the collateral damage on individual people of being constantly bombarded with environmental catastrophe is inhibiting our capacity to tackle the climate crisis.

New words to express profound feelings

The emotions people feel around the planetary crisis can be intense, life-changing, and overwhelming. Many describe being terrified, floored, or swept away by grief. "Global dread," "eco-anxiety," "environmental grief," "climate rage," "eco-paralysis," "environmental cynicism," "climate change distress"—despair about the future of the planet has garnered many labels in the research literature as academics try to understand and study the emotional and psychological complexity of our feelings about the state of the planet. Glenn Albrecht, a sustainability professor in Australia, says we simply don't have enough words to express how profoundly environmental changes affect us. He has created a new lexicon of terms, including *solastalgia*—the feeling of homesickness we experience when we are still in the same place, but it has been irrevocably changed. Solastalgia is distress caused by the transformation and degradation of one's home environment.[11]

Worries about climate change impact our most intimate decisions. A third of Americans reportedly consider climate change in their decision not to have children or to have fewer children, according to recent polls in the *New York Times* and *Business Insider*.[12] A growing number of people around the world are experiencing real anguish over whether or not to have children. They worry about the harm an additional person could do to the planet, and they feel genuine anxiety about whether a child could lead a good life on the hotter, less stable world they fear is coming.

Eco-anxiety is overwhelming kids

It's not just adults who are suffering. In our noble zeal to emphasize the urgency and enormity of environmental issues, we appear to be inadvertently raising a generation that feels hopeless about the future of the planet. A 2018 international review of recent research on the psychological impacts of climate change on children published in *Current Psychiatry Reports* reveals that many kids honestly believe the world may end during their lifetime as a result of climate change or other global threats. In-depth interviews with ten-to-twelve-year-olds in the US found that 82 percent of children expressed strong feelings of fear, sadness, and anger when discussing environmental problems.[13]

I want to underscore that these are kids who have *not* directly experienced catastrophic floods, droughts, sea level rise, or bushfires. These findings are from researchers who specifically study the psychological impact of indirect or gradual climate change effects.

The reason I think that is such a sad and important point is that children are suffering emotional and psychological anguish not from their lived experience, but as a result of their *anticipation* of a dystopian future they believe is inevitable. They see planetary destruction as a foregone conclusion. They are so deeply embroiled in the narrative of doom and gloom that they have no idea other futures are possible.

It's not surprising they feel this way. They are growing up in a media storm of end-of-the-world threats and getting graded on homework assignments that hammer home the magnitude of environmental problems. This pervasive and skewed orientation toward analyzing what's broken follows them throughout their school careers.

Though it's natural and responsible to try to protect the people we love by focusing on the dangers they may face, lots of studies now show what parents and teachers already know. The best way to equip kids to handle challenges in their lives is to help them learn how to develop and maintain strong social networks within supportive communities. They also need to learn how to develop effective, creative problem-solving abilities to overcome adversity. These same strategies are true for the challenges they may face from climate change. The Australian Psychological Society provides more details on how to do this in a helpful online guide called "Raising Children to Thrive in a Climate Changed World." In it, they remind parents: to talk about but not catastrophize the problem of climate change; to validate and help children learn to recognize their feelings; to offer emotional support; to take positive environmental action together; and to nurture kids' capacities for resilience, flexibility, and adaptability.

The more we worry, the more we ... shop?

A fatalistic focus on climate doom triggers a host of what psychologists describe as conscious and unconscious concerns about our own deaths. The result is an emotional state of "existential anxiety."

Psychologists use the term *terror management theory* to describe the constant tension each of us experiences in our day-to-day lives between our desire to live and the fact that we know that one day we will die. Without realizing it, we develop defense mechanisms to manage this psychological tension.

Fears about the death of the planet are even more visceral because we know how completely dependent our own lives are on the health of living ecosystems. Climate change is a primary driver of biodiversity loss, just as the loss of biodiversity contributes to climate change. Both of those fates—and our own—are inextricably linked.

The trouble is, depictions of climate change as an inevitable, sweaty death sentence trigger these defense mechanisms. We protect our sense of security by subconsciously denying the problem or minimizing the credibility of the threats.

So even though you might assume that people who fear death by climate change would be motivated to change their behavior, it doesn't work that way. When we already know there is a massive problem, and people just keep telling us how bad it is, we suffer real fears about our survival. In fact, fearmongering amplifies our existential anxiety, which sets off a chain of protective reactions that can cause us to downplay the issue and reduce our likelihood to take action.

Surprisingly, these fears can actually lead us to shop more. Researchers have found links between existential anxiety and hyperconsumerism. Shopping (for people who derive personal validation or identity from their stuff) decreases our sense of vulnerability.[14] In our materialistic, consumerist culture, a common response to soothe our unconscious fears of death is to hop online for some comfort shopping.

This could help to explain why Black Friday 2019 hit a record $7.4 billion in US online sales at the same time concern about climate change was at a record high. Mass consumerism is bad for the environment in a myriad of ways. Millions of shoppers buying and then discarding smartphones and TVs, for instance, contribute to the fifty million tons of e-waste the world generates each year. If you were to add up *all* the stuff people around the world consume, everything from food to birthday presents to toilet-bowl cleaner, it would total a whopping 60 percent of greenhouse gas emissions and between 50 and 80 percent of total water, land, and material use, according to a 2015 study in the *Journal of Industrial Ecology*.[15] It's shocking to realize that by slamming people with messages of climate doom, no matter how well intentioned, we may inadvertently escalate the environmentally destructive shopping we so badly need to stop.

The finite pool of worry
triggers emotional numbness

Other researchers explain our failure to act, despite high levels of concern about environmental crises, through a phenomenon known as the finite pool of worry. According to researchers at Columbia University's Center for Research on Environmental Decisions, there are limits to the number of concerns a person can deal with at one time. Overburdening people's capacity for worry with too much doom and gloom leads to emotional numbing. We tune out or feel immobilized. "When we're scared, we can freeze," says Susan M. Koger, a psychology professor at Willamette University in Oregon, who teaches and writes about psychology for sustainability.[16]

The trouble is, emotionally numb can look a lot like not caring. My friend Carrie teaches high school. She recently told me that she's been showing her classes increasingly graphic images of climate change devastation to try to shock them into caring. "They are so apathetic," she says.

Apathy can easily be mistaken for a lack of compassion, but many psychologists interpret it as quite the opposite. Apparent indifference or dissociation often serves to mask a person's feelings of helplessness.

Apathy is produced as a response to feeling powerless in the face of political realities we cannot control.[17] To avoid feeling helpless, guilty, and afraid, we create a veneer of not caring in order to maintain an image of ourselves as smart, tough, and in control. Apathy stems from fear and a lack of capacity to tackle what seems like an insurmountable task. When we believe nothing will change for the better, then any positive action can feel useless or pointless.

So if the students in Carrie's class already know about climate change (which, according to research, it's pretty well guaranteed they do), and if they keep being slammed with examples of how unjust it is or how little society is doing to correct it, her lessons may unintentionally create the apathy she is trying to cure.

There is a worrisome connection between apathy and cynicism. People who fall prey to apathy then may end up transforming their original political frustrations into longer-lasting expressions of skepticism, cynicism, and mistrust. Indeed, you don't have to look far to see this happening writ large.

A rise in cynicism and drop in trust

Pessimism and cynicism are on the rise in many countries, according to Our World in Data, a research project based at the University of Oxford that analyzes big data trends. Meanwhile, feelings of trust are plummeting. The Edelman Trust Barometer measures levels of trust in business, media, government, and nongovernmental organizations. In 2017, the barometer revealed a global implosion of trust. In nearly two-thirds of the twenty-eight countries surveyed, the general population did not trust these four social institutions to "do what is right." We're rapidly shifting from the Age of Anxiety to the Age of Cynicism.

People trust each other less in the US today than forty years ago. Indeed, the US ranked the lowest in the most recent barometer reading, positioning it as the country with the least-trusting informed public. Decline in trust between

Americans is coupled with a reduction in trust in their government, which, according to the Pew Research Center, is at historically low levels.[18] Distrust is growing fastest among younger Americans.

Not surprisingly, trust is actively undermined by fake news and gaslighting. Gaslighting is when someone manipulates the facts so often, it leaves you second-guessing your reality. It causes you to question your own judgment. If gaslighting had a mantra, it would be "repeat a lie often enough and it becomes the truth." This is especially true when the person gaslighting is in a powerful position. I imagine you can think of a few prominent politicians that fit that description. The more a gaslighter fuels our mistrust of others, the more cynical we become of other people's motives, and that spirals into pessimism, distrust, and disappointment more generally.

While intelligent skepticism is warranted—after all, one is wise to distrust untrustworthy sources—the double whammy of rising rates of kneejerk cynicism about human nature, combined with apocalyptic forecasts about the future of the planet, leaves many with the helpless feeling that the world is too broken to fix. We may become so overloaded with worries that we disconnect from the suffering of others or lose motivation to lend support, a condition psychologists call compassion fatigue. We detach, withdraw, and disengage. Fear and despair mute our ability to find creative solutions. They cause us to self-isolate. Hopelessness becomes a self-fulfilling prophecy.

Are you feeling eco-anxiety?

If you picked up this book because you are personally experiencing eco-anxiety, or climate grief, or deep worry, I hope this chapter reminds you that you aren't alone in these feelings. You aren't crazy. Lots of people feel the same way, and psychologists confirm that it makes sense. According to the Canadian Mental Health Association (CMHA), fear is a reasonable, even healthy response to the enormity and urgency of the planetary crisis.

Dr. Courtney Howard speaks on behalf of the Canadian Association of Physicians for the Environment when she says, "The intersection between the climate emergency and mental and physical health will become one of the world's major issues." The CMHA labels the climate emergency as a mental-health emergency. More than a thousand psychologists signed an open letter endorsed by the Association of Clinical Psychologists UK demanding immediate and effective action on climate change in light of the enormous mental-health impact of the climate crisis.[19] Tackling climate anxiety and tackling climate change are inextricably linked.

Your feelings are real. But the point I want to keep driving home is that they are also inflamed by a flow of information in which positive developments are almost entirely missing.

You feel deeply about the environmental crisis because you have a deep love for this magnificent planet. That love is a strong and wonderful quality, and it's empowering to find a way to excavate it from beneath all your fear and anger and grief and disappointment. When you look through the research at what triggers and sustains personal environmental behaviors, it's things like compassion rather than shaming. It's showing empathy when someone does something that they know they shouldn't do—reminding them that we're all human and mistakes are just a normal part of life. It's finding meaningful purpose in the actions. It's getting support from important relationships.

All of us experience a vast range of emotions, and it is the interplay of these feelings that enables us to move toward the world as we would wish it to be. Acceptance of *what is* is not the same as fatalism about *what comes next*. We need to be wary of seeing climate demise as a foregone conclusion. A 2018 study of fifty thousand people from forty-eight countries, reported in the journal *Climate Policy*, found that people who believe climate change is unstoppable were less likely to engage in personal behaviors or to support policies to address climate change.[20] Conversely, according to a 2014 study by leading climate change communication researchers, when someone understands that climate change is a truly dire problem *and* they have a sense of the effectiveness and feasibility of the ways people are collectively acting to solve it, then they are more likely to take action themselves. Recognizing both the threat and the potential solvability of the climate crisis is paramount to mobilizing action.[21]

Fatalistic forecasts are also being co-opted and used for ulterior motives. Climate doom, according to Michael Mann, distinguished professor of atmospheric science at Penn State, is the new climate war—and it's just as dangerous as the old one, which focused on the denial of the science. If people aren't causing climate change, as the deniers purport, then there is no reason to limit the use of fossil fuels or make the societal-level transformations necessary to achieve a zero-carbon economy. Likewise, if it's already too late to make change or we're past the point of no return, as those expressing climate doom say, then there's no point to policy reform or for the largest emitters to change their ways. In a 2019 interview for the *Guardian*, Mann says that propagating frightening environmental narratives "leads people down a path of despair and hopelessness and finally inaction, which actually leads us to the same place as outright climate-change denialism."[22]

Dire predictions and apocalyptic claims from the past are also being used to undermine the need for urgent climate action now. President Trump did just that when he dismissed environmental activists as fearmongering "prophets of doom" in his 2020 speech to the World Economic Forum: "They predicted an overpopulation crisis in the 1960s, mass starvation in the 70s, and an end of oil in the 1990s," he said. "These alarmists always demand the same thing: absolute power to dominate, transform, and control every aspect of our lives. We will never let radical socialists destroy our economy, wreck our country or eradicate our liberty."[23] It's a clear example of how alarmist rhetoric can backfire and give ammunition to deniers.

WHEN YOU CHOOSE to reject doom and gloom you are caring for your own well-being—and standing up for badly needed changes. As we'll see in the next chapter, hope is not complacent. It is a powerful political act.

3

HOPE IS CONTAGIOUS

My daughter, Esmé,
was my last good egg.
To my utter surprise, I went into menopause
immediately following her birth.

On Midway Atoll
Wisdom, a Laysan albatross,
is over sixty years old.
She continues to lay
one good egg, each year,
from a nest
amid the plastic of the Great Pacific Garbage Patch.

Against all odds, good things come from good eggs.

ONE OF THE things I didn't expect when I started talking about hope was that people would go out of their way to tell me how wrecked the world is. As soon as I mention a positive emerging environmental trend, someone invariably tells me about the worst possible environmental situation they can think of. It's as if they think I don't know what is happening to the planet.

I do know. I am the same age as the modern environmental movement, which began in the US in the 1960s. The narrative of doom and gloom that dominates the way we talk about the environment is the defining story of my life. Like you, I feel devastated by announcements like the one made by the World Meteorological Organization in 2019, which stated that climate-heating greenhouse gases showed "no sign of a slowdown, let alone a decline."[1] People tend to respond to dire reports about the planet in two very different ways, according to environmental psychologist Renée Lertzman. We either see the planet as beyond rescue, or we double down on optimism. I wonder if my comments about hopeful trends unintentionally cause some people to see me as being overly optimistic, and therefore trigger them to respond with evidence for why doom is justified. But the thing is, it's not an either/or situation. What we need is a more nuanced understanding of how what we know and how we feel impact our capacity to act.

What hope is, and what it isn't

Hope is not about turning our back on the facts. It's precisely because we *do* know how much trouble we are facing that millions of people all over the globe participate in climate protests. Protests are inherently hopeful acts. Researchers who study social movements tell us that hope plays a crucial role in mobilizing individuals to take part in collective action,[2] just as participating in collective action fuels feelings of hope. Research shared by the American Psychological Association demonstrates that hope helps us to stay engaged with stressful situations, promotes coping skills, and reduces denial[3]—three important qualities given the misinformed concern that hope might breed complacency or foster climate change denial.

Ernst Bloch, the German Marxist philosopher, also saw hope as intimately entwined in the realities of everyday life and the transformative power of political action.[4] In his massive three-volume compendium entitled *The Principle of Hope*, published in the 1950s, he wrote, "[Hope] requires people who throw themselves actively into what is becoming, to which they themselves belong."

Hope is also not about feeling cheerful. When you love something and it's being destroyed, it's extremely difficult not to give up. Trying to move in a positive direction in the midst of a terrible situation takes fortitude. Being hopeful is, in many ways, the more difficult path than despair and cynicism. Hope doesn't protect you from feeling disheartened or sickened by the scale and horror of the fires in the Amazon, Australia, California, or Western Canada. Especially when the frequency and severity of "fire weather" is increasing with climate change. Feeling furious and upset at deforestation, coal-fired power plants, and politicians who fail to lead urgently needed climate reforms, or angry that you've inherited a screwed-up situation from previous generations, is justified. Outrage shows you know what's going on and you know what absolutely must change. Reaching the point of "enough is enough" spurs us to protest, boycott, and stand up for the things we love and believe in.

Anger and hope are not opposites. They have a symbiotic relationship. Both anger and hope are mobilizing emotions. The 2019 climate strikes that drew millions of people around the world to demonstrate demand for change represent a mass social movement fueled by both anger at the injustice of what is and hope for what should be. Hope is what sustains us to keep fighting for social and ecological justice.

Multiple conceptions of hope

The word "hope," for lots of people, rubs them the wrong way. It symbolizes a naive notion of thinking positively or searching for a single shiny solution to cure the vast complexity of environmental woes. I sense it's this definition of hope that Greta Thunberg railed against in her speech in Brussels on February 21, 2019, when she said: "You can't just sit around waiting for hope to come. Then you are acting like spoiled, irresponsible children."[5] The kind of hope I am talking about isn't some Pollyannaish notion of looking on the bright side. I'm not advocating that you, or anyone else, put on a smile and trust that everything will work out fine. To me, that's not actually hope. It's wishing.

Wishing and hoping are often entangled together, but they ask quite different things of us. Hopeful people, according to Shane Lopez, a cognitive psychologist and hope researcher, plan for and show up in ways they believe will help the future improve. Hope, in his definition, is active. There is an essential connection between hope and agency—our sense that we can take action. Studies show hope helps us to cope, which keeps us from disengaging from difficult situations. Wishful people, in contrast, see the positive outcome they are yearning for as out of their hands. Wishing demands no effort on our part.It's the fantasy that things will just magically turn out okay.[6]

Hope is not a new topic. Exactly what people mean by *hope* is hotly debated. No doubt people were talking about it long before written languages even existed. The priestess-poet Enheduanna wrote about hope 4,300 years ago. Pandora, the first human woman in Greek mythology, opened a jar of human evils which spread around the world. Hope, however, remained within the container. Hope was, in essence, given to us by the gods. In addition to Greek mythology, other sacred teachings position hope as located in God. Hope is central in Christianity, and the Holy Qur'ān and medieval Islamic writings have many references to "hope."[7] As God endures, so too does hope.

For others, hope is not linked to a divine being; hope resides within ourselves. Developmental psychologist Erik Erikson saw hope as a genetically encoded element within babies that is activated by the nurturing, positive interactions we experience as infants. Hope is innate, he said, but it can fail to be fully actualized if this compassionate care is missing. Even in adulthood, hope, according to Erikson, requires ongoing attention and restoration. By delivering compassionate care, physicians, he argued, could restore hope to their patients.[8]

Philosophers, spiritual leaders, psychologists, poets, neuroscientists, and others continue to try to tease out the essential elements of hope. Lately there's been a heightened interest in hope in the academic literature. Some say it's a response to a growing collective sense of crisis, and the lack of political and ideological direction to solve it. The recent Hope & Optimism project, developed by Notre Dame, Cornell, and the University of Pennsylvania, assembled a massive collection of research papers probing the theoretical, empirical, and practical dimensions of hope, as well as seeding artistic explorations of hope through screenwriting and playwriting competitions.

With such a long and diverse history, it's not surprising many different ideas about hope exist. Hope is often conceptualized as a belief in the possibility of a better future. It's also seen as a state of mind, as a disposition or capacity, as an emotion, as a process or activity, and often as a combination of all of these things. Peter Burke, emeritus professor of cultural history at the University of Cambridge, introduces the notion of scale into the conceptualization of hope: we have *big hopes*, such as hopes for a better world, and *small hopes*, including hopes for a better life for ourselves or our families.[9]

Hope, it turns out, is contagious. We often don't even realize when we are picking up someone else's feelings. We spread despair or hope to our colleagues, our families, and our friends, and we catch emotions in return. According to a massive 2014 study using Facebook, hope, despair, and other emotions are contagious online too.[10] Researchers manipulated the amount of emotional content in the news feeds of almost 700,000 Facebook users. When people received more negative content, they produced more negative posts. The opposite pattern occurred when they received more positive content. We catch and spread emotions on Facebook just as we do through our real-world interactions.

Rather than holding a single definition of hope—a one-size-fits-all approach—I find it useful to see hope in terms of what Anthony Wrigley, a philosopher of ethics at Keele University, calls a "cluster concept," whereby what we mean by "hope" varies depending upon the context. Different conceptions of hope for different circumstances feels like a better fit for the vast and varied scope of environmental issues with which we must contend.

So, what kinds of hope do we need to carry us through the complicated global transformations the planetary crisis demands? It's helpful to think in terms of three broad categories of situations in which we find ourselves seeking hope: 1. Hope to drive our individual environmental actions; 2. Hope to power societal transformations; and 3. Hope in the face of hopelessness.

1. Hope to drive our individual environmental actions

Individual environmental actions, even when they feel way too small, matter far more than we might realize. A growing number of studies critically analyze which individual actions have the greatest collective positive environmental impact. Top of the list? Live car free, avoid air travel, eat a plant-based diet, don't waste food, ditch single-use plastic, and push for Earth-friendly policies.[11] So how do we find the hope we need to make these impactful changes in our everyday lives?

Hope researcher Shane Lopez spent many years collaborating with his mentor, Rick Snyder, a pioneer in the scientific study of hope. Cognition is the mental act of knowing. Hope, they argued, is not simply an emotion, but rather a cognitive process—a way of thinking—that influences our emotions. Hope, they said, is something we actively create when we have goals we care deeply about, real plans to reach them, and the motivation and ability to accomplish them. We can *think* our way into feeling hopeful.

You've probably heard the expression "where there's a will, there's a way." Rick famously used it to frame his goal-directed concept of hope (which he called "hope theory"). Hope requires both the *will* (the motivation and ability to make something happen) and the *way* (plans and strategies to pursue those desired goals). Hope, from this perspective, requires us to believe that our actions might help to change things for the better, and also to believe we have the capacity to plan and carry out those actions.

Hope theory is anchored in the belief that human actions are goal-directed. It assumes that people have enduring beliefs about their abilities to achieve their goals. "High hope" people bounce back when things go wrong. They reassess their goals, modify plans, or pursue slightly different outcomes that have more doable paths. As we know all too well, the sorts of circumstances that require us to hold onto hope are never easy; if there were simple answers, we would have already found them. "High hope" people expect things to be difficult and they expect they'll be able to rise above them.

Hope, within the cognitive framing of hope theory, is a personal attribute that can be measured, increased, and deployed. Rick and his collaborators created different hope scales, which continue to be widely used in health, education, business, and other settings to measure the extent to which a person expresses both the "willpower" and the "waypower" to solve problems.

It's valuable to identify concrete steps one can take to feel hopeful. However, there's something about the assumption that hope can be wrestled into a multi-point scale, and that we can be scored against it, that I find troubling.

How does hope theory apply to people when they are feeling too hopeless to see any pathways forward? When I look into the face of someone who believes the planet is screwed, talking about individual actions can feel trivial, simplistic, and out of scale with the enormity, urgency, and complexity of global issues. And if hope is a rational goal-setting enterprise, what befalls those who are too ill or cognitively impaired to process rational thoughts? Are they in essence *beyond hope*? Surely that isn't true.

And what about the prevalence, within hope theory, of labelling people who score highly on these tests as "high hope" people?[12] Can we really classify someone as a "high hope" person? Would I characterize myself that way? Rather than a fixed identity, hope, for me, is context dependent. I feel super hopeful about some things and despondent about others. And my feelings, and sense of agency and self-efficacy, shift and change in response to all kinds of influences, internal and external. I suspect it's the same for many, if not most people. Even the most positive, goal-directed, capable folks I know regularly experience bouts of futility, discouragement, and helplessness.

The notion of measurable, attainable, self-propelled hope embedded within hope theory is so logical, it's easy not to notice how much power it ascribes to us as individuals. It's predicated on the assumption that if you want something badly enough and can think through the right route to achieving it, even though the outcome is not certain, it could actually come true. I question the way hope theory positions individuals as omnipotent.

Yet even with these reservations, I find hope theory compelling, and useful when it comes to taking personal environmental actions. I find it reassuring to imagine that hope is within our own purview: it depends on our *personal* motivations, pathways, sense of agency, and goals.

Through their research, Rick Snyder and his collaborators found that hopeful people had the ability to find many routes to a goal. They also found people could be taught to develop this skill. In other words, we could shape our hopeful futures—and we could be taught how to do so.

So how can you learn to be hopeful? According to hope theory scholars, when you find something you are deeply excited about, *and* you are good at it, invest your time in doing it. You will start having a flurry of hopeful thoughts. Hopeful thinking causes hopeful emotions, in the sense that positive feelings emerge from our sense of our own capacity to achieve our desired goals. Feeling hopeful about one thing carries over to other situations, hope theory says. Successes help you to feel excited and enthusiastic, and thus more hopeful about other things.

How individuals help bold actions become everyday practices

Plus, there's a shortcut to hopefulness. Choosing to spend time with a hopeful person enables you to witness how that person puts those steps into action, making it easier for you to apply them to your own life.

Recently, I heard about research done by a doctoral student named Steve Westlake. He found that people who take bold environmental action, like giving up flying, can have a wider knock-on effect (what psychologists call "social influence") by helping to shift what other people view as "normal."[13] I got to witness a great example of this firsthand through my friend Sarah-Mae Nelson. I met Sarah-Mae several years ago in Monterey, California, when she decided to try to live for a year without using plastic. She gave herself particular exceptions—she could use plastic if she was at the hospital for medical treatment, for instance—and she was fully aware of the plastic in the paint on the walls in her apartment and at work, and so on; things that were beyond her control. But for the things she could control, she fully committed to the challenge. Even products like canned tomatoes and milk in glass bottles turned out to be tricky, and she found herself petitioning companies to remove the plastic lining in canned foods and the plastic bottle caps on milk that comes in glass bottles.

Two specific things stuck with me in her approach. First, for every piece of single-use plastic she used, even if she specifically told someone not to give her a plastic straw or to place a plastic sticker on a bagless apple, she would do fifteen minutes of beach cleanup. She held herself accountable as an individual, even if the plastic was thrust upon her by someone else. The second thing was that she lived with a roommate who continued to bring home lettuce and other produce in single-use plastic bags. Sarah-Mae did not get sidetracked by what her housemate was doing. She focused on her own actions. She didn't demand others live plastic free; she just did her best to do it in her own life and shared "how-to's" with anyone who wanted to know more about it. She may not have influenced her roommate (I don't know what that person is doing now), but I witnessed the impact of her bold action as it spread to many other people.

Some people argue against focusing on individual climate change actions. They worry that it distracts us, shifting our attention away from the companies most responsible for emissions. And, they have a point. According to researchers at the Climate Accountability Institute, a mere twenty companies are responsible for 35 percent of the world's energy-related carbon dioxide and methane emissions.[14]

But according to a 2019 report from the Grantham Research Institute at the London School of Economics and Political Science, keeping our focus on major players *is* happening: lawsuits against governments and the highest greenhouse-gas-emitting companies are being brought by investors, activists, shareholders, cities, and states in at least twenty-eight countries around the world. Holding those in power legally accountable for climate change has become a global phenomenon.[15] We need to be wary of those who use people's worry and guilt as a ploy to shift attention onto their individual actions and away from fossil fuel industries. Yet that doesn't mean we should shy away from the huge contribution individual changes do make.

As long as we recognize there are political actors who would rather we *only* focused on individual actions, then we can hold them to account *and* get on with recognizing the astonishing collective value of individual action. Indeed, I bet you've already changed lots of things in your everyday life to benefit the planet and your health and well-being. And I am sure there are people in your life, like Sarah-Mae, who inspire you with their bold quests. Maybe *you* are that person.

What's exciting is that bold quests don't stay bold for long. They become ordinary as more and more people embrace doing them, and that's a super hopeful thing to watch happen. According to behavioral psychologist Kelly Fielding of the University of Queensland, Australia, we are very influenced by what other people do, even though we may not think we are. Humans are highly social beings. We look to others to see how we should behave. As more and more of us take positive climate action, more and more of us follow that example. Seeing evidence that what each of us does makes a difference motivates us. It's hopeful, and that hope is empowering.

2. Hope to power societal transformations

Tackling the climate crisis and biodiversity loss demands societal transformations of epic proportions. When social transformation is the goal, hope is a collective action. Duane Bidwell of the Claremont School of Theology describes how hope emerges through interactions with the people who surround us.[16] Creating positive action as part of a community instills hope within each of us.

To hope is not to wait around until you are feeling optimistic, but to join with others in defiant response to what we are doing to the planet. It is an *action that you do* rather than a *feeling that you have*. "Hope is a verb with its sleeves rolled up," David Orr, a leading environmental thinker, famously said back in 2008 in a conversation reported in the *Earth Island Journal*.[17]

Former UN climate chief Christiana Figueres argues that the only way we can save the planet is with relentless, stubborn optimism. We must actively decide to remain hopeful. As writer and political activist Barbara Kingsolver says, hope is "not a state of mind but something we actually do with our hearts and our hands, to navigate ourselves through the difficult passages."[18]

Ibram X. Kendi makes a parallel argument in his book *How to Be an Antiracist*. Whether or not someone is antiracist, he says, has little to do with self-proclaimed identity. It doesn't matter whether we claim to be antiracist. What matters is what we do. We either act in a way that contributes to inequality, or we actively try to dismantle inequality. Any of us, at any time, according to Ibram, can fall into either category, depending on what we support or fail to support.[19]

Inequity and environmental destruction are intimately linked. Social injustice drives climate change, which in turn drives further inequity. According to a 2019 study by earth scientists at Stanford University, the difference between groups of nations with the highest and lowest economic output per person is 25 percent greater than it would have been without climate change.[20] The climate crisis is the culmination of a long history of exploitation of people and other species. It is fueled by an unsustainable economic system that privileges endless growth over justice and equity.

Transformative changes demand that we take stock and critically analyze underlying systems and conditions that keep us hurtling toward a rapidly warming planet. We need to recognize social injustice and privilege and how our own actions or inactions make us complicit within these systems. The deeper we dig to understand the root causes of our collective failure to live sustainably on Earth, the more overwhelmed we can become. Feelings of shame, guilt, and futility spiral into emotional paralysis and cynicism, especially if we believe that we'll never be able to change things in time.

If you are feeling this way, believe me, you aren't alone. In schools and universities, we actively teach people to critically analyze the environmental crisis, which is a wise and important thing to do, yet too often we fail to support the feelings such critiques engender. The result is a teeter-totter of emotions that finds students and their teachers seesawing between momentary highs of hope and downward crashes of despair. Despite our best intentions, critique becomes a disempowering habit. As Maria Popova, the brilliant curator of *Brain Pickings*, puts it, "Critical thinking without hope is cynicism. But hope without critical thinking is naiveté. I try to live in this place between the two."[21]

What's missing is an essential middle step. This middle step is where creativity, innovation, and imaginative visioning happens. We need to nurture the capacity to imagine something beyond what is.[22] When we remind ourselves not only of what is broken but also of specific responses to issues that are yielding positive results, we are more apt to see how we can contribute our own creativity and energy to this positive momentum. "Hope," says Henry Giroux of McMaster University, "makes the leap for us between *critical education*, which tells us what must be changed; *political agency*, which gives us the means to make change; and the *concrete struggles* through which change happens."[23]

Go ahead and express those "outlaw" emotions

As eco-anxiety, climate despair, and other emotional responses to the planetary crisis become more common, sharing these difficult feelings with others is ever more important. Yet as Lisa Kretz, a professor of philosophy at the University of Evansville, says, Western cultures all too often undermine, devalue, and dismiss the attempts of groups to give voice to the sadness, grief, and despair they feel about the destruction of Earth's inhabitants. The silencing of these "outlaw" emotions, as Lisa describes them, impedes those grappling with ecological loss from recognizing their shared feelings and channeling that energy into social change.

This is a problem with both individual and societal implications. Burying our feelings or trying to ignore our emotions is bad for our health. According to a 2019 study in the *International Journal of Psychotherapy Practice and Research*, more than 80 percent of visits to the doctor have to do with a social and emotional challenge, while only 16 percent have a strictly physical cause.[24] People who have difficulty expressing and managing their emotions are more likely to suffer from unhealthful behaviors, like poor nutrition, lack of exercise, abnormal sleep patterns, and substance abuse. Bottling up their feelings in turn creates stress and tension, which can exacerbate other health problems.

Conversely, those who find ways to express their emotions reap significant benefits. In fact, a healthy relationship with one's emotions is a more significant predictor of longevity than factors such as diet or activeness. People who have a sense of hopefulness and can deal with sadness by finding purpose and meaning are more likely to live longer and healthier lives than folks who tend toward pessimism.

Collective hope

Maria Ojala, a psychologist at Örebro University in Sweden, found that kids too were more able to cope with their worries about climate change when they were able to give voice to their climate-related fears and hopes, put their trust in their communities, and see themselves as part of a larger movement. According to Maria, "coping is not only an individual act for children but also part of a social process within the adult world."[25] Many children worry that fixing the world rests on their shoulders. A decade ago, one of my former graduate students, Carly Armstrong, found that when it comes to coping with climate change, kids need to express their feelings and to see themselves as part of a larger community response of people working together to solve these important issues.[26]

The most startling thing about disaster, writes Rebecca Solnit in *A Paradise Built in Hell: The Extraordinary Communities That Arise in Disaster*, is not that people rise to the occasion, but that so many of them do so joyfully. She found what many hospice workers and chaplains know: that the despair of crises is often intertwined with the joys of community, meaningfulness, and clarity of purpose.

It's encouraging to watch the rise of people bravely sharing how they're feeling about the planetary crisis. Growing numbers of people in the United States, Australia, Britain, and Canada are actively seeking out climate anxiety support groups, according to a 2019 Reuters news report.[27] The "Stand Up for Climate Change" initiative uses sketch comedy, stand-up, and improvisation to encourage people to engage with their feelings through humor.[28]

People looking for ways to share their feelings about the planet can also take inspiration from creative projects that invite people to gather around food to discuss difficult issues and strengthen supportive relationships. Projects include the Restaurant of Mistaken Orders in Japan, where all of the servers are people living with dementia, or Dinner Over Death, which invites folks to host a meal with friends and family to discuss end-of-life decisions. A group of students in Montreal created Vent Over Tea, which offers free, in-person and confidential listening sessions in local cafés to promote mental wellness and connection.

Having other people validate and echo the legitimate sadness and mourning we feel helps us to build emotional solidarity, says Lisa Kretz. Mourning in isolation can spiral into immobilizing despair, whereas shared grief builds communities of support that can help people move toward more energizing emotions of hope and anger. "Hope," she says, "can function as a bridge from mourning to action."[29]

A rising tide of activism

Social justice activists, climate action groups, and anti-capitalism protesters took up the Indigenous rights cause of the Wet'suwet'en hereditary chiefs in February 2020, as part of a broader fight against resource-extraction projects in Canada. The groundswell of mutual aid in the form of hot food, borrowed tents, donations, music, art, medical services, and social media updates surrounding protesters at the British Columbia legislature building was so abundant that the protesters shared the food donations made to them with the Mustard Seed food bank. That same month, Teck Resources Limited withdrew its application to build the $20-billion Frontier oil sands mine in Northern Alberta, citing Canada's divisive debate over climate change.[30] It's a decision, some analysts say, that reflects the drop in oil prices as the global energy sector moves away from fossil fuels and toward zero carbon.

The year 2020 began in the same fashion that 2019 ended, in a sea of protests that surged across six continents. According to Erica Chenoweth, a political scientist at Harvard, more people in more countries were involved in nonviolent mass movements in 2019 than any time in recorded history.[31]

With literally millions of people taking to the streets in 2019, civil resistance brought down leaders in Algeria, Bolivia, Iraq, Lebanon, and Sudan. They continue to push against power in Ecuador, Egypt, Georgia, Haiti, Peru, Poland,

Russia, Iran, and Zimbabwe. Controversial policies in Chile, France, and China were reversed. Protests are ranging across Latin America, Eastern Europe, Russia, Canada, the US, and a number of African and Asian nations. The #MeToo movement has provoked questioning of gender relationships around the world.[32]

In response to mounting anger and frustration with social injustice, inequality, economic disparity, and inaction on the climate emergency, ordinary people are coming together in unprecedented numbers and are actively changing global political cultures. Many of them are young. Almost 42 percent of the world's population is twenty-five years of age or under. In Asia and Latin America (where 65 percent of the world's people live), a quarter of the population is under fifteen, and in Africa, that figure rises to 41 percent.[33] Young people are rising up against extreme social and political inequities to fight together for justice and equality in numbers never before seen. The year 2019 will be remembered as the year youth-driven climate justice marches spread around the world. Those marches sparked so many climate emergency declarations that by the end of that same year, one in ten people on the entire planet were living in a place that had committed to decreasing greenhouse gas emissions.[34]

3. Hope in the face of hopelessness

When someone feels overwhelmed with eco-anxiety or hopelessness, we need to listen. We need to empathize with the depths of futility, shame, remorse, guilt, or sorrow they are feeling; to let them know that it is okay to sit with grief and accept whatever emotions emerge. It is not helpful to try to suppress our sadness. If we acknowledge the full range of our feelings, eventually hope can return.

Yet creating safe spaces for people to express their environmental grief is something we rarely do. We lack any recognized infrastructure to support children, or adults, suffering from despair about the planetary crisis. It's as if we are engaged in a mass movement of emotional denial.

Not so long ago, the same thing was true for the way people treated the end of life. Throughout the US, Canada, the UK, and much of the rest of the Western world, technological advances in the 1950s meant not only that more people were being treated in hospital but also that more people were dying in hospitals rather than in their homes. Death, in this context, was viewed as a medical failure, which contributed to the prolonging of life in hospital settings beyond the point that many would argue was meaningful existence. People knew about the suffering of the terminally ill and dying, yet those feelings were not considered to be the concern of health care institutions.

In 1963, Dame Cicely Saunders, a British nurse, social worker, and doctor, introduced the idea in a lecture at Yale University that the dying needed and deserved specialist care. She wrote and lectured widely, pushing back against the overmedicalization of death and prioritizing comfort and emotional support over disruptive medical interventions.

Cicely created St. Christopher's Hospice in South London in 1967. It quickly became a model for the ways in which evidence-based pain management and compassionate care could contribute dignity and respect to dying people.

Essentially, she launched the modern palliative care and hospice movement. Hospices all over the world embrace the spirit of Cicely's words: "You matter because you are you, and you matter to the end of your life. We will do all we can not only to help you to die peacefully, but also to live until you die."[35]

The hospice movement has much to teach those struggling with hopelessness in the midst of a planetary crisis. Hope is most often conceived of as a "future" orientation. We hope for a better future. Yet palliative care specialists tell us that within end-of-life situations, when death is inevitable, hope still exists. That's because hope is also grounded in a sense of a meaningful present. We hope that the life we have led and the time we have at this very moment matters to ourselves and the people we love. Hope, thus, tells us what we value. When we share our hopes, we create a mutuality of values that supports and strengthens our relationships with others who care about the same things we do.

Conceptualizing hope as an innate quality that lives within us regardless of the situation is a different perspective to that espoused by Erik Erickson, who saw hope as something that a physician could give or take away. Writing in the *Journal of Clinical Oncology,* near the time of his own death, Peter Yuichi Clark, director of Spiritual Care Services for UCSF Health at the University of California, San Francisco, cautioned against equating *hope* in a medical context with *hope for a cure.* To do so would mean that those diagnosed as terminally ill would necessarily be without hope. It is not up to our doctors or the people who love us to worry about whether to "offer false hope," he said, because hope is not theirs to give.[36] The very notion of false hope is faulty. It is based on an assumption that hope, authentic or false, is something that can be given. But hope is not externally produced. It is intimately tied to one's personal sense of meaning. *Engaged hope*, Peter said, is a form of realistic hope in which we discern how to live with intention and integrity even when we are ill, trusting that somehow our actions will benefit others and make a difference.

Václav Havel, the Czech statesman, also saw hope as an orientation of one's spirit: a state of the mind and heart rather than a state of the world. Hope, he said, is about finding meaning regardless of how apparently hopeless the circumstances. In *Disturbing the Peace*, he wrote: "Hope... is not the conviction that something will turn out well, but the certainty that something makes sense, regardless of how it turns out."[37] As the celebrated Austrian psychiatrist and Holocaust survivor Viktor Frankl stated in *Man's Search for Meaning*: "We must never forget that we may also find meaning in life even when confronted with a hopeless situation."[38]

Hope demands that we look truth in the eye

These definitions of hope urge us to look deeply and realistically at things as they actually are. Hope is possible in the face of climate change, not because we pretend the complex issues impacting the planet don't exist, but because we fully embrace the truth of the situation.

Embracing the truth, unfortunately, is difficult to do. As I shared in chapter two, almost all the news we hear about the environment is bad, and we rarely hear much detail about what's working. That makes it almost impossible to get a complete picture of what's deteriorating *and* what's improving at any given moment. Without a full picture, we easily fall prey to assuming the worst.

Coming to grips with things is also challenging because we don't have much control over when we hear environmental news or how much we can handle at a particular time. When a patient gets a frightening diagnosis, that person, ideally, gets some say in how much they want to know and at what stage they are ready to hear it. Some thought goes into the emotional impact of the bad news that is being delivered, and increasingly, health care professionals receive training through programs like VitalTalk[39] to help them tailor these difficult conversations to individual circumstances.

We don't do anything like this when it comes to talking about the environment. We throw down statistics about the shocking rate of species extinctions or the amount of plastic in our food as if they carry no emotional implications whatsoever. As a case in point, I was recently contacted by a journalist who wanted to know the best way to talk with kids about climate change. The question was sparked by a recent incident at a Toronto school. An eight-year-old child had come home visibly upset. "We're all going to die in eight years," she told her mother. The girl had attended a climate change presentation at school that featured Greta Thunberg's emotional speech to the UN Climate Action Summit in New York in 2019: "You have stolen my dreams and my childhood," Greta said. "People are suffering. People are dying. Entire ecosystems are collapsing. We are in the beginning of a mass extinction, and all you can talk about is money and fairy tales of eternal economic growth." After watching a video of the speech, the children were shown the climate change doomsday clock[40] as it rapidly counts down the years, days, and seconds we have left. Imagine being a kid seeing this and believing we are literally moments from the end of the world.

When teachers and other trusted adults take on the role of telling kids just how wrecked the world is, in the name of *telling them the facts so they'll be inspired to act,* they fuel the cycle of fear, anxiety, and hopelessness I described in chapter two. Too often, we launch into doomsday scenarios and unintentionally leave children hanging. Without a hopeful framing around these dire issues, we reinforce an immobilizing culture of fear from which hopelessness emerges as a profound threat. It leaves us all caught, as Joan Halifax, an activist, anthropologist, and hospice caregiver, says, "in a collective imagination of horror that is, like a tsunami, drowning hope."[41]

Hearing a diagnosis of a terminal illness is devastating. Hearing a diagnosis of a dying planet is devastating. We know compassionate ways of handling life-altering circumstances, and we need to apply that same thoughtfulness to the way we talk about threats facing the planet.

You never know what is going to happen

Sometimes, the best you can do when you truly feel hopeless about the planet is to try to remember none of us knows how things are going to play out. We live in a world that is alive with 8.7 million other kinds of species. The sheer amount of biodiversity means it's impossible to know the outcome of the combined actions of all of the living beings on the planet. We have science-informed models to help us predict and ward off critical problems, and it's essential that we respond to them with all the seriousness and urgency they deserve. Yet it's also true that even the best projections exist within a context of unknowable variation.

Wise hope, says Joan Halifax, comes from accepting that we *never* know what is going to happen. "Wise hope is born of radical uncertainty, rooted in the unknown and the unknowable," she says.[42]

To hope is to accept the best and the worst of a situation. Dreadful things happen. People we love die. Species go extinct. Feelings of hopelessness touch all of us. We all suffer. Yet even in dire circumstances, things can change in ways we never expected. An interview with Vince Crisostomo on CBC Radio marking the thirtieth anniversary of World AIDS Day provides a poignant reminder.[43] Vince was diagnosed with HIV in 1989. His doctor literally told him he had "six months at best." Now, three decades later, Vince is almost sixty. No one could have foretold in the early days of HIV/AIDS that, thirty years later, a modern HIV diagnosis is no longer a death sentence. Vince courageously described the confusion and bewilderment that you feel when death is imminent—and then you don't die. He regrets passing on the chance to buy a house for very little money. He didn't join his company's retirement plan. He made no plans for his future because he was told he didn't have one.

We must look at the planetary crisis as realistically as humanly possible, and then open ourselves to be surprised. We must act with our best intentions and efforts, in response to the most accurate knowledge, all the while accepting that we cannot know what exactly will unfold, now or in the future. When we accept that we truly don't know what will happen, we also accept that possibility exists. Hope exists in the possibility of transformation.

4

STORIES CHANGE

First came mosses,
then trees,
then insects,
and all the birds and mammals
that feast upon them.

In just thirty-three years
there is a forest
of ten million trees
covering the 230 square miles
of Washington State

that the volcanic eruption of Mount St. Helens
sent hurling
into the sky.

Y EARS AGO, I visited my friend Leslie at her desk in a television newsroom. This was back in the days when people still took pictures using film, and I noticed a series of faces pinned onto the tack board against her office wall. "People who've been murdered," she told me, explaining how she kept them handy so she could easily find them when a newsworthy moment associated with the crime arose and she needed a visual. I've often thought how soul-sucking it must have been to share an office with those images.

Yet gruesome pictures have now become a normal part of all of our lives. Every time we glance at our phones to check the news, we open ourselves to a deluge of images: runaway bushfires, bleached coral reefs, skylines obliterated by burning coal, mountains of waste, a dead whale with a stomach full of plastic. These images lodge in our heads, and they become the stories we tell ourselves about the state of the planet. The circumstances they depict are so awful it doesn't even occur to us that those situations could change.

When the news brings peak moments of horror, give yourself time to feel your feelings of grief—and then keep track of what comes next with this story. Look for what happens *after* the worst. As we discussed in previous chapters, environmental news is heavily focused on problems, so we have to actively search for hopeful developments to get a fuller, more accurate sense of real-time changes.

I keep an image on my laptop to remind me to do this. It is a colorful coral atoll in an azure-blue tropical ocean. The photo was taken in 2017 and it exudes glorious nature. But it's the location where it was taken that makes it special: Bikini Atoll, the world's largest testing site for atomic bombs, is now a thriving coral kingdom.

Today, if you slipped into the waters off Bikini in the tropical Pacific, you'd find corals ablaze with colorful fish and large tuna, snapper, and sharks. It is so beautiful that you might not recognize its historic identity. It provided the inspiration for the famous bathing suit. And it's where the US Army detonated the first H-bomb. Twenty-three nuclear explosions were carried out at Bikini between 1946 and 1958. The cost to people and the environment is incalculable. It condemned the Islanders who lived there to a life of perpetual displacement.[1] It was the most violent series of events ever perpetrated on the ocean.

Yet, sixty years later, diving along the edge of the mile-wide Bravo Crater—created in 1954 by a nuclear blast one thousand times stronger than the bomb dropped on Hiroshima—Stanford University biology professor Stephen Palumbi and his team found corals as big as cars in healthy coral communities. Some of these corals may have begun growing just ten years after the last bombs were dropped.[2]

Bikini is damaged *and* thriving. It is radioactive and resilient. Acknowledging the restoration doesn't diminish the horror. I am humbled by the idea of healing happening on

such a vast scale and in such extraordinarily unlikely circumstances. Even in nuclear disaster, there is the potential
for rebirth.

Indeed, gray wolves flourish within the Chernobyl Exclu-
sion Zone (CEZ), the vast area surrounding the remains of the
Chernobyl Nuclear Power Plant, which exploded on April 25,
1986. The wolves are radioactive—and thriving. The population density of wolves within the zone is estimated to be up
to seven times greater than in surrounding reserves, according to Michael Byrne, a wildlife ecologist at the University of
Missouri. At 1,660 square miles (4,330 sq km), the CEZ is one
of the largest wild sanctuaries in Europe.[3]

Bikini and Chernobyl remind me that life is far more resilient than I ever expected, and that environmental situations
are rife with ambiguity. Young mountain gorillas in Rwanda
have taught themselves to disable poachers' snares.[4] Sudbury,
Ontario, one of the most polluted landscapes on Earth, is
now an international model of ecological recovery. Environmental situations are not all bad or all good; they demand
nuanced understandings. The crooked spider web glistening
in a Chernobyl forest, like the bald head of my dear friend
Katie, after chemo, is imbued with the capacity to be broken,
beautiful, or both.

Remarkable resilience

Nature has an astonishing capacity for healing. Not just the
kind of healing it brings to our human hearts, but the capacity of ecosystems to recover. I think those of us who work
with the environment are reluctant to talk about this other

kind of healing, for fear it might be co-opted to justify the appalling degradation of the planet. "Don't worry about PCBs, my friend. The Earth will heal itself." That sort of thing. How can we speak of species flourishing in the midst of nuclear disaster, or the capacity of wetlands to decontaminate themselves from oil spills, without opening the floodgates to further environmental degradation? If nature can heal itself, the argument goes, why do we need to change?

We know we do have to change. The landmark 2018 report by the UN Intergovernmental Panel on Climate Change stating we have only twelve years to limit climate change catastrophe is just one of a legion of highly respected reports providing more than ample evidence of this. Celebrating resilience in no way diminishes the urgent need for environmental reforms. It stokes the courage we need to keep making the massive necessary changes.

It's a mistake to focus only upon fragility. Things get horribly broken. That is true. But the remarkable capacity for renewal is true as well. The recovery of devastated places reminds us that life is complicated. Life is marked by complex mixes of opposing forces: destruction and restoration; intention and outcome; hope and despair. Gloom and doom are intimately balled up with possibility. Far from making us complacent, stories of resilience and recovery in remarkable circumstances fuel hope and counteract the cynicism and emotional paralysis I discussed earlier.

I am in awe of the ability of stories to shape our lives. Anthony Giddens, the British sociologist, talks about the dualism of structure and agency—how the grand narratives that tell the story of what it means to be a woman or an activist or a scientist influence and are in turn influenced by the personal stories we tell about and to ourselves.

Once, at a UN meeting, I met a foreign aid worker who was kidnapped on a road in Cameroon. "I knew I was dead," he said. "So, I changed the story. I thanked them for picking me up. I told them how grateful I was that they had rescued me on such a dangerous road. I made them my rescuers. I stuck with it no matter what they did to me." After several terrible days, he was released.

Steve Johnson, a community organizer, told me how Portland, Oregon, came to be such a beacon for urban agriculture and sustainable living. "Back in the seventies, a bunch of us used to drive around in vans with signs on the side, campaigning for environmental issues. We wanted to make Portland a place where people would come for the green lifestyle. We published articles about this amazing city. Whenever someone would write or call and say, 'Is it really that green?' we'd always say, 'Yes.' It wasn't, of course, but we figured if they got here and they didn't like what they saw, they'd leave. But if they got here and were still committed, those were the kinds of people we'd need to make the story come true." Today Portland is one of the greenest cities in the US.

Remember, life constantly changes

Life on Earth constantly changes, often in ways we never would have guessed possible. Yet with environmental issues, there is a real tendency to think that situations are stuck, or always on a downward trajectory from utopia to ruin. We conflate change with assumptions of inevitable destruction.

The result is that we're often working with out-of-date assumptions that are just plain wrong. This is a problem because there's a close relationship between hope and accurate information. Studies reveal that when we don't know about past improvements, we are much more likely to be pessimistic about the future.[5]

Take violent crime, for example. Surveys show that most Americans think violent crime is getting worse, and they think it has been getting worse for a long time.[6] In fact, according to the FBI's annual reports and annual surveys by the Bureau of Justice Statistics, violent crime in the US has fallen sharply since the early 1990s. Violent crimes still occur, of course, but overall their incidence is way down and continuing downward. But that's not what most people believe.

That's because most of us develop our perspectives about what's happening in the world by generalizing our personal experiences rather than checking current statistical facts. Because positive trends often unfold slowly, they don't get much attention, so we rarely hear about them. Instead, we are influenced by newsworthy events that highlight the unusual. We end up unintentionally carting around a bunch of outdated perceptions and heightened fears and using them to support our beliefs. The result is that we hold inaccurate assumptions about situations changing for the worse, say researchers at the Ignorance Project—an initiative of Gapminder, an independent Swedish foundation that uses reliable statistics to promote a fact-based worldview.[7]

If we don't think to look for change, we fail to see the shifts that are occurring all around us. We miss the fact that giant pandas are no longer endangered. Or that in fifty years, bald eagles have gone from the brink of extinction to numbers never before seen. Bald eagles are so commonly spotted around Victoria, BC, where I often spend time, that many people no longer remark when they see one.

We may miss dramatic changes in the relationships we have with other species. In the span of a single generation, dogs have become family members. This change in sentiment gave birth to pet-adoption programs, which, combined with spay-and-neuter initiatives, have successfully lowered the euthanasia rates of dogs and cats in the US by more than 92 percent since 1984.[8] Americans spend more on their pets than ever before—roughly $95.7 billion in 2019.[9]

Meanwhile, sales of plants are booming, especially among people under the age of thirty-five. One in three American households grows food at home or in a community garden, a 200 percent rise in the past decade.[10] Two-thirds of Londoners bought a houseplant in 2019. Those in the know say the wild popularity of plants is driven by a desire to green urban living spaces, to nurture a living being, to connect with nature, and to find refuge among plants from the stresses of climate change and technology.[11]

We need to situate our understanding of what's happening with the environment in real time and real circumstances. The trouble is, we rarely do this. Instead, we rely on decades-old slogans like "Save the Whales," because we assume whales always need saving. But not all whales do. The population of South Atlantic humpback whales, for example, jumped from 440 individuals in the late 1950s to a 2019 count of 25,000, which is near their historic population size before commercial whale hunting almost drove them extinct.[12]

Generalized slogans trap us in a fixed state of discouragement. Tigers, for example, have been "on the brink of extinction" for at least forty years. I used to say those exact words when I was a summer interpreter at the Calgary Zoo in the early 1980s. Do a Google search right now and you'll likely see tigers described using this same phrase. But can something that lasts almost half a century really be described as a "brink"? The statement lacks specificity and context, and without these vital criteria the information tells us nothing, and only serves to make us think the situation is hopeless, or makes us more cynical about environmentalists' claims.

The truth is, different populations of tigers live in different parts of the world. They are all endangered. Yet the population of Amur tigers (or Siberian tigers, as they are also called) is currently making a positive comeback in Russia. In the 1930s fewer than 20 individuals remained. That number climbed closer to 540 by 2018, according to Russia's Natural Resources and Environment Ministry. This positive trajectory owes much to President Putin's affection for photo ops with Amur tigers, and his government's funding of tiger conservation and research projects. And to the Spatial Monitoring and Reporting Tool (SMART), which is being used in Russia, and in countries all over the world, to detect and monitor threats and provide real-time information to wildlife law enforcement teams.[13] That's vitally important because poaching of tigers—and the prey they depend on—is a major threat.

Russian conservation teams, made up of more than one hundred full-time members, share data and prioritize responses using SMART as they patrol key tiger habitats in armored personnel carriers.[14] SMART also enables data sharing and improved collaboration across the larger tiger conservation community, and across borders. Efforts to recover Amur tiger populations extend across the Russia/ China border, with several regional protected areas recently designated to improve habitat connectivity between the two countries. A vast territory used by Amur tigers and Amur leopards along the Russian and North Korean borders is part of a new national park system.[15]

Amur tigers are endangered, *and* their population is increasing in Russia. Both statements are true. It requires a shift from "either/or" thinking to "both/and" thinking. Knowing how dire the circumstances are *and* what's working to improve them is essential for identifying solutions, like cross-border protected areas and better monitoring through SMART, that can be tailored to benefit other conservation efforts.

Crossing bridges

A critical issue for Amur tigers is that individual tigers have difficulty getting to each other to breed because their wild habitat is broken up by a massive highway. Grizzly bears face a similar challenge of fragmented habitat, even in Banff National Park, where one of Canada's busiest highways runs straight through the park.

One solution, in both cases, is to build safe ways for them to cross those roads. Indeed, the world's first tiger underpass opened in Russia in 2016.[16] No doubt the political will to build it came from the success of wildlife bridges and underpasses springing up all over the world. Wildlife bridges are just one piece of a massive movement to create wildlife corridors that join together wild and protected lands.

Structures to help animals cross busy roads come in many forms, depending on the species involved. A crab bridge across a highway on Christmas Island, Australia, helps crabs return to the sea. Engineers from the West Japan Railway Company designed U-shaped tunnels to enable turtles to safely cross under the tracks. For shyer animals that prefer dark tunnels, there are underpasses, like those used by pumas in Brazil or water voles in London, England. Six hundred wildlife bridges, called ecoducts, enable badgers, bison, elk, and wild boar—and sometimes even hikers—to cross highways safely in the Netherlands.

Crossing structures save the lives of wildlife, and people. Arizona uses fencing and military-grade target-acquisition software to detect and activate flashing lights that alert drivers in real time when an elk is near the roadside. The result is a 97 percent reduction in animal-vehicle collisions.[17] In the Mojave Desert, tortoise roadkill dropped 93 percent after fences were placed on either side of a highway, and turtle tunnels were installed.[18]

Banff started experimenting with wildlife bridges and underpasses back in 1996. As of 2019, six wildlife bridges and thirty-eight underpasses exist along a fifty-five mile (90 km) stretch of highway.[19] It's the most extensive system of wildlife crossings in the world. Wildlife deaths have been reduced by 80 percent across species. For deer alone, the deaths have dropped by 96 percent.[20]

In a twist on the age-old question of *Why did the chicken cross the road?* researchers in Banff used infrared cameras to monitor how wildlife use highway overpasses and underpasses. Grizzlies, it turns out, took five years before they started using overpasses on a regular basis, whereas elk are less discerning and started using them as soon as they were built. Black bears and cougars prefer to use underpass tunnels, whereas grizzlies, deer, moose, and elk prefer the open sightlines offered by the overpasses. Such preferences make sense when you consider that black bears and cougars normally live in woods, while grizzlies and ungulates are more at home in open meadows.[21]

Wildlife bridges, underpasses, and corridors create opportunities for animals to move out of broken-up "islands" of habitat and to interact with a greater gene pool. Genetic tests on more than ten thousand hair samples from bears using the crossings in Banff confirm that providing ways for bears to reach mates with different genetic backgrounds is working. The tests reveal that the crossing system is helping to maintain genetically healthy populations of grizzlies and black bears.[22]

Joining up fragmented habitats and supporting migration routes of animals big and small is happening all over the planet. Marigolds, sunflowers, and other nectar-rich flowers are popping up across balconies, cemeteries, schoolyards, and industrial sites as people in Oslo, Norway, collectively create a bee-friendly corridor to help pollinators cross the city. In Toronto, the land beneath hydroelectric towers is being restored to butterfly meadows. An eight-story vertical butterfly meadow forms the facade of a new building in Manhattan. And the largest property owners in London's West End are working together to lure wildlife back into the heart of the city by creating the Wild West End—a green corridor that joins up existing parkland using green walls, planters, green roofs, flowerboxes, and pop-up spaces.

A green corridor on a much grander scale, the RIMBA corridor, connects existing national parks and sanctuaries in Sumatra, Indonesia. The result is a network of wild habitat for orangutans, Sumatran tigers, and other endangered populations. By enlarging green spaces that provide vital ecosystem services, the RIMBA corridor also benefits people directly—bringing cleaner water, better retention of soil nutrients, and increased carbon storage.[23]

Rewilding

Green corridors and wildlife bridges are part of a rapidly growing rewilding movement, designed to introduce native wildlife back into degraded ecosystems to regenerate wildness and natural processes. The goal is to increase biodiversity, enhance greenhouse gas sequestration, and improve people's access to nature.

Witnessing the return of giant anteaters to the Iberá Wetlands of northeastern Argentina,[24] wild bison to the mountains of southwest Romania (and Banff National Park), and wolf packs to Yellowstone Park in the US inspires hope. As species recover, they drive habitat changes that help to restore ecosystems.

You can now hike seventeen hundred miles across seventeen national parks and more than sixty Patagonian communities in Chile, thanks in part to the world's largest donation of private land. Kristine McDivitt Tompkins made the donation in 2019 on behalf of Tompkins Conservation, the foundation she created with her late husband, Doug. (He was the founder of clothing brands Esprit and North Face; she is the former CEO of the clothing brand Patagonia.) Over the past three decades, the Tompkinses initiated conservation projects in Chile and Argentina that have added more than eleven million acres (about 45,000 sq km) to the national park systems of those countries. By purchasing overgrazed sheep ranchland, removing livestock and fences, working with volunteers, communities, and experts to restore habitat, and reintroducing wild native species, they are nurturing biodiversity on a grand scale.[25]

Natural climate solutions

These and a myriad of other examples remind us that successful restoration practices already exist. Some are being scaled to a global level. In 2011, the government of Germany and the International Union for Conservation of Nature (IUCN) launched the Bonn Challenge, a global goal to restore 150 million hectares (370 million acres) of degraded and deforested land by 2020. As of January 2020, enough land had been pledged to exceed that goal by 15 percent. The Bonn Challenge now calls for 350 million hectares (865 million acres) to be under restoration by 2030. It's just one of many exciting examples of what Greta Thunberg is calling for when she says we need to "protect, restore and fund natural climate solutions."[26]

Natural climate solutions occur when we conserve and restore ecosystems—and improve land management— through actions that increase carbon storage or avoid greenhouse gas emissions. They include approaches like rewilding, which uses habitat protection and restoration to benefit biodiversity and decrease emissions. Combating habitat loss and species extinctions is good for the climate, and vice versa.[27] Indeed, in 2017, a team of scientists found that natural climate solutions could get us at least 30 percent of the way to the greenhouse gas reductions set by the Paris Agreement, *and* provide many more biodiversity benefits than other climate change approaches.[28]

It's thrilling to see how quickly life returns. In the largest dam-removal project in the history of the United States, the Elwha River now runs freely from a snowfield in the mountains of Washington's Olympic National Park to the Pacific Ocean. Salmon started to return to their natal waters upstream almost immediately after the dams were removed in 2014.[29] Reservoir beds that looked like moonscapes now host vibrant young forests and wetlands where elk graze.[30] The return of beavers to the Elwha watershed is a boon for salmon. Beavers drag branches, making shallow water channels where juvenile salmon can safely travel, and beaver dams create slower water habitats where the insects that salmon feed upon thrive.[31]

Watching rivers spring back to life is infectious. In 2019, a successful crowdfunding campaign in the Ukrainian part of the Danube saw the removal of ten small dams. Not only will removing the dams improve water flow and revitalize local vegetation, which will serve as fish spawning grounds, it will also revive riverside meadows, which will enhance local cattle grazing and reduce flooding of agricultural lands and the local highway.[32]

One of the world's most famous rivers, the River Thames in London, England, was declared biologically dead in 1957. Today it is home to 125 species of fish, and 138 harbor seal pups, who were born in the river estuary in 2018. More than 3,000 harbor and gray seals now live in the river,[33] along with harbor porpoises and sometimes dolphins and whales. The Thames is now said to be the cleanest river in the world that flows through a major city. This remarkable comeback from the dead owes much to environmental protections and regulations that have reduced the flow of pesticides and fertilizers into the river, as well as to happenstance. Pollution from toxic metals has dropped since 2000, in part due to the fact that as people switched to digital photography, silver—a common pollutant associated with film cameras—diminished.[34]

The Cuyahoga River in Cleveland, Ohio, used to be so polluted it caught fire—thirteen times. The last fire, in 1969, came just a few months after another environmental disaster, a catastrophic oil spill off the coast of Santa Barbara, California, that released three million gallons (11 million liters) of oil into the ocean, killing thousands of birds, fish, and marine mammals. The sight of a burning river, coming so soon on the heels of the image of wild birds drenched in oil, sparked outrage that drove positive change. The first Earth Day occurred ten months later. Within three years, the US Congress had created the Environmental Protection Agency and passed the Clean Water Act.[35]

Stories change. Restoration is possible. Our attitudes change, too, even toward species we most fear.

Making friends with sharks

Twenty years ago, when I first moved to Monterey, California, a sighting of a great white shark in Monterey Bay made the news. But over two decades something remarkable has been happening. Monterey Bay is undergoing an astonishing recovery, and as that beautiful ecosystem grows healthier, top predators, like white sharks, are returning.

Monterey is the place John Steinbeck wrote about in *Cannery Row*: a fish-canning town that all but disappeared when the sardines were fished out. The water was so polluted, people called it an "industrial hellhole." The sea otters and whales, once plentiful, had been hunted to near extinction. It was an environmental catastrophe.

Today it's a world-class center of ocean conservation. Fifty marine research institutes and organizations, including the Monterey Bay Aquarium and Stanford University's Hopkins Marine Station, cluster around the Monterey Bay National Marine Sanctuary.[36] It's the ideal place to demonstrate how new policies and those put in place decades back combine to create very significant impacts. Monterey Bay is healthier than it has been in the past two hundred years. Sea otters have returned. Humpback whales are becoming year-round residents. Wildlife watching is so reliably great, the BBC chose to base *Big Blue Live*, the very first television series showing wildlife events happening in real time, in Monterey Bay.

Not surprisingly, the number of people surfing, boogie boarding, stand-up paddleboarding, diving, kayaking, and swimming in those same waters is also on the rise.

So, what do you think happens when more white sharks and more people use the same waters?

More shark attacks, right?

Wrong. The likelihood of being bitten by a white shark dropped 91 percent between 1959 and 2013 despite the tripling of California's coastal population, which now exceeds 21 million people.[37]

Thanks to the US Marine Mammal Protection Act, which came into law in 1972, there are more northern elephant seals, more harbor seals, more California sea lions—more animals that white sharks prefer to eat. The recovery of marine mammals has been dramatic. (In the 1800s they were hunted to near extinction for their blubber, which was used for lamp oil.) The total population of northern elephant seals in the entire North Pacific Ocean had dwindled to fewer than forty individuals by the end of the nineteenth century. Sixty years ago, there were no elephant seals in the Año Nuevo State Park Natural Preserve, off the coast of Santa Cruz, California, where white sharks now feast on them. But if you visit Año Nuevo this January, when migration is in full swing, you'll find more than 3,500 elephant seals congregated at just this one park. The northern elephant seal population in the North Pacific Ocean is now estimated to be 170,000 individuals.[38]

A hopeful change in public attitudes toward sharks is under way in coastal California, and around the globe. In 2019, Canada became the first country in the world to ban the import and export of shark fins. In the past decade, seventeen shark sanctuaries have been established around the world, driving positive press about sharks and the countries, like Palau, that protect them.[39] And the digital revolution is changing the way scientists study sharks, enabling people, for the first time in history, to see the world from a shark's point of view.

The digital conservation revolution

Barbara Block and her team at Stanford University's Hopkins Marine Station began using biologging tags on white sharks in 2000 to follow their migrations and behaviors. Like wearable technology for fish, these sophisticated tags collect and send information from individual sharks that researchers can track on their iPhones and other personal devices.

Tracking devices create a combination of specificity and personalization that drives marine conservation. They disabuse us of the notion that the ocean is one big bathtub full of water, too vast and nebulous to protect. From a great white shark's perspective, the Pacific is an assemblage of seasonal hotspots and well-defined highways. Tagged sharks reveal precise places that matter most to them.

Along the coast of California, for example, data from the Stanford lab shows that each September, subadult and adult great white sharks start arriving in Tomales Bay, north of San Francisco; the Farallon Islands, about thirty miles (50 km) from the Golden Gate Bridge; and Año Nuevo (the elephant seal beaches), just north of Santa Cruz. They stick around these coastal hotspots throughout the fall and early winter, much of the time within waters protected by California's network of marine reserves.

Come late winter that changes. Most of the more than two hundred great white sharks tagged off the central coast of California eventually make their way to an open ocean region halfway between Hawai'i and California. Researchers dubbed it the White Shark Café in recognition of its popularity. These hotspots turn out to be essential, not only for great white sharks but for a host of other ocean predators too. The findings of a decade-long project published in 2011 that tracked twenty-three species of top marine predators reveal remarkable overlap in the migrations and habitats of blue whales, elephant seals, leatherback sea turtles, and several different species of tuna, sharks, and seabirds. Witnessing life in the Pacific at this level of specificity enables researchers to better predict when and where individual species are likely to be and to provide policy makers with data to protect vitally important habitats.

In March 2015, the National Oceanic and Atmospheric Administration announced that it would expand the size of both the Gulf of the Farallones and the Cordell Bank National Marine Sanctuary, where one of the planet's most significant populations of white sharks, and so many other highly migratory marine predators, return each year. This ocean region is so rich in wildlife that Barbara Block calls it the "Blue Serengeti." Like its African namesake, the region hosts diverse and rare species. More than twenty-five endangered or threatened species, thirty-six marine mammal species, and over a quarter million breeding seabirds depend on these waters. The expansion of the sanctuaries, coupled with the existing network of state marine reserves, makes the California coast one of the world's most protected marine environments.

The images and mapping data captured on the tags also create unprecedented opportunities for anybody with a cell phone to forge a personal relationship with a white shark. The Shark Net app invites people to see the places where white sharks live and to get to know them as individuals, just like friends on Facebook or Google+. Individual sharks are identified by notches and other markings along the trailing edge of their iconic dorsal fin, as well as by marks and scars on their bodies. Some have been given names, like Scar Girl, Flat Top, or Tom Johnson, and given that white sharks can live to be seventy years or older, there is a wealth of stories to learn about each of them. Created by the Stanford University research team, the app provides detailed personal information about these individual sharks as well as real-time notifications when one of them is detected by an acoustic receiver. Scientists say the intimacy generated through the app enables people to form attachments to individual animals, which aids shark conservation. The more we understand the natural behavior of sharks, according to research studies, the less we fear them and the more we support efforts to conserve them.[40]

Welcome back whales

If we don't look to see if stories have changed, we miss huge things, like the impressive recovery of the California blue whale population—which is back to being almost the size it was before commercial whaling began.

This is fantastic news for Monterey Bay and the other marine areas these whales frequent. It's fantastic news for all of us. People used to think that if the great whales returned, then all the fish would get eaten. It isn't true. Whales, it turns out, actually create the conditions that help fish to thrive. Whales often feed at great depths, and when they return to the surface to breathe, they churn up the water column, spreading plankton and nutrients. They may migrate long distances to mate, bringing nutrients with them to far latitudes where the water has fewer nutrients. Whales also produce vast amounts of poop, rich in iron and nitrogen, which effectively fertilizes microscopic plants called phytoplankton, upon which tiny marine animals feed. Fish, in turn, feed on these small creatures as well as the plants. More whales means *more* fish.[41]

And that's not all. In 2019, researchers declared large baleen whales to be the "carbon-capture titans of the animal world."[42] A single large baleen whale absorbs an average of thirty-three tons of CO_2 throughout its lifetime. Part of their carbon capture capacity is due to that phytoplankton I just mentioned. Whales increase phytoplankton productivity, and phytoplankton plays an enormous role in atmospheric conditions. Phytoplankton produces two-thirds of the planet's oxygen. It captures 40 percent of all CO_2 produced. According to researchers, the ocean's phytoplankton captures as much CO_2 as 1.7 million trees. That's *four* Amazon rainforests' worth.

If you like economic arguments, here's a beauty. If you add up the contribution a single whale makes to carbon capture, the fishing industry, and the whale-watching economy, a single whale is worth two million dollars. That makes the global population of whales worth more than one trillion dollars.[43]

This is a hopeful equation, because the number of whales on Earth is increasing. A staggering 2.9 million whales died

due to commercial whaling between 1900 and 1999. Remarkably, some populations of gray and humpback whales have
almost returned to their pre-hunting numbers. Fin whales
have been upgraded from endangered to vulnerable, thanks
to conservation efforts.[44] Though many populations of sei
whales and blue whales remain endangered, their numbers
are on the rise too. Today scientists estimate that 1.3 million
whales live on Earth. It's thrilling to chart the rise in that
number across the years. It's believed four to five million
whales once lived on Earth. If they were to return to those
numbers, whales could capture 1.7 billion tons of CO_2 annually while continuing to improve the health of the oceans
and fisheries.[45]

The rise of marine protected areas

In 2000, only 0.7 percent of the world's oceans were designated as a marine protected area (MPA). Nearly a decade later,
in 2008, the Global Ocean Legacy project hired me to write
the scientific brief to try to convince then president George
W. Bush to establish the world's largest MPA in the Marianas
Trench, which is within US territorial waters. Thanks to all
kinds of good work by lots of people, the project succeeded.
The Marianas Trench Marine National Monument was formally designated on January 6, 2009.

It is no longer the world's largest MPA. It's not even close.
In just ten years, it has been surpassed over and over and
over again, by the establishment of new, bigger, ecosystem-
scaled areas of protected ocean. In an inspiring example of
international cooperation, twenty-four countries and the

European Union created the world's largest marine sanctuary in Antarctica's Ross Sea in 2016. It covers an area of ocean
larger than the entire country of Mexico. As of 2020, it is the
biggest protected area on the planet.[46]

The establishment of MPAS is just one step in a long process of truly protecting the oceans. As the writer Brian
Payton wisely points out, "protection of wildlife and wilder-
ness is not an achievement but an ongoing, intergenerational
project."[47] As new threats arise and old problems persist, protected areas demand constant vigilance. Still, designating
nearly 8 percent of the ocean is cause for celebration. That's
more than a tenfold increase, much of it within just the past
few years. It's a hopeful reminder of how quickly change
can happen.

More exciting still, marine protected areas with the
highest levels of protection are reversing degradation and
rebuilding the resilience of ocean life. When we get out of
the way, other species flourish. The biomass of fish in marine
reserves is on average 670 percent greater than in adjacent
unprotected areas, according to a 2017 meta-analysis across
MPAS.[48] In some MPAS, scientists report that there are more
fish, and bigger fish, in some cases within just three to five
years of a reserve being protected. Marine protected areas
also support more complex ecosystems that are more resilient to the effects of climate change than unprotected areas.
And though they are intended to conserve wildlife within
their boundaries, marine protected areas create a spillover
effect, enhancing local fisheries and creating jobs through
ecotourism.

A second-Earth mentality

I began this chapter talking about how images shape the stories we tell ourselves about the planet. We need to be mindful of the fact that while images are frozen in time, stories are always changing. We need to be aware that the preponderance of media we see about the environment focuses on problems, and actively look for what happens after the worst.

A few years ago, I had the great fortune of spending time at the Rachel Carson Center for Environment and Society (RCC) in Munich, Germany—an international think tank with a focus on environmental history. It was there that I developed a deeper appreciation for the power of images, even those we may never have seen ourselves but that nonetheless shape the way we live. One day, as I was wandering the halls, I bumped into Donald Worster, an eminent environmental historian and professor emeritus at the University of Kansas. Don led me into his RCC office, where he showed me an image of a remarkable map created by Rumold Mercator in 1587. The map showed the "new" global perspective that was emerging from the fresh data gathered by Christopher Columbus, John Cabot, Ferdinand Magellan, and others during the Age of Exploration.

Maps will always be hotly disputed, value-laden depictions of the world. In addition to its blatant colonial perspective, what was so prophetic about the map from an environmental perspective, according to Don, was the way it depicts the world not as a single globe but rather as two coequal spheres. One sphere or planet contains the "old world" of Europe, Africa, China, and so on, and the other planet contains a "new world of the Americas." "It is a map of Second Earth," Don told me. In the minds of the Europeans at that time, this second Earth promised a massive windfall of infinite opportunity, a whole new world's worth of natural resources. The map is a symbolic representation of the beginning of the Age of Abundance, which fueled the rise of capitalism and the idea of limitless growth that has driven development for the past five hundred years.[49]

A clear indicator that we still carry the vestiges of a second-Earth mentality is Earth Overshoot Day. Calculated annually by the Global Footprint Network, it's the day each year that we begin to use more natural resources than the Earth can produce in a year. In 2019, we consumed a year's worth of resources in just seven months. We see the evidence of this overconsumption in hard-hitting reports, like the May 2019 Global Assessment Report on Biodiversity and Ecosystem Services, which shows that nearly one million species are at risk of becoming extinct.[50] Mass extinction of the natural world ranks with climate change as the greatest threats we face.

Nature needs half

Today another image sticks in my mind alongside the Second Earth map. It is the Half-Earth map; a high-resolution, dynamic world map based on the geospatial location and distribution of the world's species. It highlights areas of special importance for biodiversity protection and habitat restoration.[51] What it tells us is that to safeguard the bulk of biodiversity, reverse the species extinction crisis, and ensure the long-term health and stability of the planet, nature needs half of the Earth's land and oceans.

How to accomplish this is what the Global Deal for Nature is all about. It's a science-driven plan with real timelines to conserve the diversity and abundance of life on Earth, and it's rapidly gaining high support. In 2019, business networks in twenty-four countries representing two thousand companies pledged immediate action to halt biodiversity loss.[52]

Currently, a little less than half of the planet remains in its natural or semi-natural state. In 2010, governments around the world committed to global targets to protect nature under the United Nations Convention on Biological Diversity. They agreed to set aside 17 percent of land area under protected areas and 10 percent of marine and coastal areas. The good news is they came close to achieving those targets, but there is now widespread scientific consensus that it's not enough. When world leaders meet in 2020, the goal is to raise the targets to protect 30 percent of the Earth by 2030. We also need to restore or maintain an additional 20 percent of the planet in its natural state to act as climate stabilization areas (areas such as forests that have a disproportionately high capacity to capture carbon from the atmosphere). By combining the Paris Agreement on climate change with the Global Deal for Nature, we have a clear, evidence-based pathway for action.[53]

Today, nearly a quarter of the land on Earth, and 80 percent of its terrestrial biodiversity, sits within the territories of the world's 370 million Indigenous people.[54] These lands hold at least 22 percent of the carbon stored in tropical and subtropical forests, making them globally important carbon sinks.[55] In 2019, the Intergovernmental Panel on Climate Change added international support to secure community land rights to fight climate change.

"The biggest opportunity for large scale conservation, if we are talking about hectares and intactness, it is all located on Indigenous lands," says Valérie Courtois, director of the Indigenous Leadership Initiative. "If Canada is going to meet its target of protecting 25 percent by 2025, it isn't going to happen by buying up small tracts of private lands that are available in the south. It's by negotiating with nations in the north, saying, *Is this something that you are interested in?* And, the majority of the answer is *Yes, we've been asking for this for generations.*"

Valérie, and the organization she leads, are part of a rapidly growing national and international movement to conserve Canada's boreal forest—a globally significant carbon bank, home to a quarter of the world's unfrozen fresh water, and the planet's most intact forest ecosystem. "The boreal forest is an Indigenous forest," she says. "There are fifty-one first nations in Canada which are distributed among 634 individual communities. The vast majority of those communities and bands are located in the boreal forest. A mark of success of those cultures has been to not only live in that ecosystem but to protect it and care for it in order for it to be pristine. To meet its conservation targets, Canada has to look to the boreal."

Mapping Indigenous territories

I wrote much of this book in the traditional territory of the W̱SÁNEĆ Peoples. I know this because Eric Pelkey,

Community Engagement Coordinator for the WSÁNEĆ Leadership Council, confirmed the location; because my kids are privileged to be part of a school district that works closely with Indigenous elders and community workers; and because of Native Land, an Indigenous-led, crowdsourced, interactive website that is populating a map of traditional territories of Indigenous Peoples, treaties, and languages.

At any given moment, you can check where you are in relation to Indigenous communities and treaty settlement lands across Canada that are now included on Google Maps and Google Earth. You can consult LandMark, an interactive global platform that maps lands collectively held and used by Indigenous Peoples and local communities around the world. The International Union for Conservation of Nature (IUCN), the world's largest environmental network, now hosts a registry of territories and areas conserved by Indigenous Peoples, where communities themselves provide case studies, maps, stories, photos, and data.

These maps remind us of what environmental historian Richard Grove calls "green imperialism" and the colonial mentality of creating parks and other protected areas by removing Indigenous Peoples from the land.[56] "Canada is a colonization country," says Valérie. "We have a history in this country of conservation being used as a reason for extirpation of Indigenous Peoples from the land." She points to the history of dispossession in Canada's oldest national parks, including Banff and Jasper. According to RadLab, an activist research collaborative based at the University of Manitoba, "high-profile mainstream environmental movements in Canada and the U.S. have often inadequately addressed or acknowledged this colonial and racialized history and analysis."[57]

Indigenous-led conservation

Amid the rising call to decolonize conservation and increase Indigenous-led practices, Canada's newest national park represents a positive step. Thaidene Nëné prioritizes traditional territory management, essentially inverting the power structure between the Łutsel K'e Dene First Nation and the federal government. Three different governments are working together to manage these 10,241 square miles (26,524 sq km) of boreal forest and barren lands—part territorial park, part wildlife conservation area, part national park reserve.[58]

In 2018, Canada designated its first modern Indigenous Protected Area. Located in the ancestral territories and waters of the Dehcho Dene People, Edéhzhíe Protected Area is a spiritual place with tremendous cultural and ecological significance.[59] Twice the size of Banff National Park, Edéhzhíe represents an important shift in recognizing and supporting Indigenous leadership and self-determination with respect to protected areas and conservation.

Indeed, in the twenty years that Valérie has been working in conservation in Canada, she says that 85 percent of proposals to establish new protected areas have been led or co-led by Indigenous Peoples. "If you look at national parks alone, the new additions all have some form of co-management agreement or process with Indigenous Peoples," she says. The same trend is true in Australia, where Aboriginal and Torres Strait Islander people manage more than half of Australia's protected area network on land.[60] The May 2019 Global Assessment Report highlighting the horrifying rate and worldwide scale of biodiversity loss included a notable exception—the rate of biodiversity loss is generally reduced on lands managed by Indigenous Peoples and local communities.

Indigenous-led conservation respects the relationships that exist between people and other species—interconnections that span millennia. It values traditional knowledge that has been passed down orally for hundreds of generations. In 2016, in a vivid example of Western science recognizing the positive impact human occupation has had on ecosystems, Andrew Trant and his co-researchers wrote in *Nature Communications*, "Human occupation is usually associated with degraded landscapes but thirteen thousand years of repeated occupation by British Columbia's coastal First Nations has had the opposite effect, enhancing temperate rainforest productivity."[61] Western red cedar trees, in particular, were found to be stronger, taller, thicker, and healthier when they grew on shell middens—vast assemblages of clam and other invertebrate shells that have accumulated over thousands of years. Clamshells release calcium and lower the acidity of the soil as they break down in middens. Clam gardens, some dating back 3,500 years, have been cultivated by Indigenous Peoples along the west coast of Canada.[62] The shells continue to help the forest grow bigger.

What if rivers had rights?

Strengthening Indigenous and community land rights is a critical human rights issue, and it's crucial for solving the climate crisis and conserving biodiversity.[63] Prioritizing traditional ecological knowledge is changing the way judicial systems relate to the other-than-human world. In 2017, the Whanganui Iwi (the local Māori tribe of Whanganui) won recognition for the Whanganui River to be granted the same legal rights as a human being, with human guardians appointed to protect its interests. After 140 years of negotiation, the victory marks a turning point in recognizing the Māori worldview of the river as ancestor and living entity.

Ecuador has enshrined the rights of nature "to exist, persist, maintain and regenerate its vital cycles" in its constitution.[64] Nature holds the status of a "juridical entity" in Bolivia, in recognition of its value to the collective public interest.[65] In 2017, in northern India, parts of the Ganges and Yamuna rivers were granted rights of personhood in an effort to combat pollution. In 2019, four years after a massive algal bloom cut off drinking water to half a million people, the city of Toledo, Ohio, granted Lake Erie legal personhood. Just as parents are legal guardians of their children, the people of Toledo now hold legal guardianship of the lake, and the right to sue polluters to pay for cleanup costs. Valérie Courtois tells me that the people of Déline are exploring the possibility of granting the rights of a person to Great Bear Lake, a lake in the boreal forest the size of Belgium.

Such new legal developments signal the awakening of a re-evaluation of the place of humans within the global ecosystem. They challenge the "humans at the top" regime that views nature as property. With our sophisticated understanding of ecological systems, it's well past time for us to unshackle ourselves from the Great Chain of Being that has elevated humans above other species since the days of ancient Greece.

We can choose the stories we live

Olle Carlsson, a Swedish polar explorer, fell in love with Antarctica years before he ever visited there. He fell in love. It just happened. And because he was in love, he learned everything he could about the place. He spent years interviewing people who had been to Antarctica, reading old logs from polar expeditions, and reading scientific journal articles. He set out to write a book himself, reasoning that by doing so, he would get the chance to go there. He arrived in Antarctica and fell even more deeply in love.

Antarctica is the only place that everyone on the planet is legally entitled to visit. You don't need a passport. No country owns it. It is governed by a commission of twenty-eight countries. When Olle learned of plans to allow mineral exploration, he went to the Swedish environment minister, who knew him well, and asked what Sweden's position was on the issue. "We think we're going to take a regulatory approach," he said. "The US and UK are pro, so we'll use that to try to slow down the rate of exploitation." Olle thought for a moment. "We could do that," he said. "But I have another suggestion. What if we advocate to do nothing?"

In international-policy speak, that's called a moratorium. You decide that no one will take any action for a certain period of time. The international whaling commission is the most familiar example. It started when whales became commercially extinct several decades ago. Eighty-nine signatory countries agreed not to resume the hunt until whale populations, "stocks," as they call them, have returned to at least 54 percent of their historic levels. In the ensuing decades something remarkable happened. A lot of people fell in love with whales. What began as a fisheries agreement to recover commercial whaling has evolved into an international force for whale conservation.

Because Olle fell in love, he dared to ask for the image of the world as he wanted it to be. He asked for Antarctica to remain free of commercial mining. In the end, that's the position the Swedish government took, and the one that was ultimately accepted. Today, Antarctica can only be used for peaceful purposes.

WE CAN CHOOSE the stories we live. We are shaped by the stories we tell.

Hope lies in the capacity of stories to transform. As we'll see in the next chapter, the health, social justice, and environmental values of Gen Z are radically altering what we eat for the better—and driving changes in how we engage in Earth-friendly living through personal technology.

5

THE AGE OF
PERSONALIZATION

You make the clouds
as you breathe
deeply in—and out.

Each breath,
a gift
of water droplets
dust
microscopic fungi

the seeds
of cloud formation.

HALF OF THE world's population is under thirty years old. Youth represent the largest demographic of people on Earth. They are the first generation to grow up with an awareness of climate change, and they are committed to tackling the climate crisis. It's a concern that crosses all socio-economic sectors and national boundaries. A 2019 survey of Gen Z across every continent included twenty countries, classified by their level of economic development: either developed, emerging, or frontier. Young people in frontier countries, including Jordan, Kenya, and Nigeria, expressed the most concern about climate change and the highest commitment to creating a sustainable future, followed closely by those in emerging economies.[1]

Gen Z want, expect, and demand that the way they live their lives be planet-friendly and sustainable. The environment influences how they see themselves and what they look for in others. The majority of people on the dating site Tinder, for instance, are under age twenty-five. They are more likely than older users to mention a cause they are passionate about on their profile, and what they're most passionate about is climate change and the environment, according to Tinder's "2019 Year in Swipe." One of the most popular emojis on

Tinder in 2019 was the facepalm, which was used 41 percent more than in the year before. It's a graphic example of Gen Z's shared sense of shock and disbelief at what's happening in the world.

If you're in this age group you know the environment is no longer a movement, it's an embodied value, a lived ethic. That shift is changing how the planet eats. It's changing what corporations do. It's changing how cities look and function. Tackling climate change demands that we transform the whole economy, that we change the way we live, the way we move, and how we spend money. This youthful global community is hyperconnected, and it's using its collective power to do just these things to create eco-friendly change on a massive scale. It's already transforming the world, and it's doing it by making everything personal.

We are living in the Age of Personalization. The marriage of digital technology and personal devices—the meteoric growth of smartphones and social media—drives our expectation for personalized experiences. Personalization became so ubiquitous so fast, it's no longer noteworthy that Netflix keeps track of our individual movie preferences. We simply expect Google Maps to remember our favorite routes, and takeout places to prompt our regular order. We depend on Fitbits to monitor our heartbeats, Siri to answer our questions, and Pandora to recommend music we might like. *Make it real. Make it immediate. Make it tailored to me. Give me real-time feedback. Show me the collective impact. Create personalized experiences and deliver them in an authentic way so I feel valued, appreciated, heard, and involved.* These are the expectations we now bring to how we work, learn, socialize, and care for one another.

Real-time feedback
improves environmental behavior

For many years, environmental educators struggled with the challenge of trying to make big, complex issues like climate change relatable to people's ordinary lives. "If only we could show them in real time how their individual actions impact the bigger picture," they would say. A plethora of ways now exist to do exactly that, and they're yielding exciting results. In a 2019 study, for instance, researchers installed smart shower meters into hotel bathroom shower stalls. The meters provided real-time feedback on how much energy the guest was consuming while showering.

What's exciting about this study is that it was conducted in real hotels with real hotel guests who did not know they were part of a study. They did not self-select to be part of an energy-conservation initiative and they didn't receive any financial reward or hotel discount for lowering their energy use, and yet they still behaved in a more environmentally friendly way. Hotel guests who received real-time feedback used 11.4 percent less energy per shower than hotel guests who didn't have a smart meter. They voluntarily lowered their resource use when they received personalized feedback on their consumption as it was happening.[2]

Many utility companies now provide their customers with real-time readings of how much energy they consume in their homes. When programs combine this kind of person-alized feedback with messages that are tailored to a specific community of people, good things happen. My friend and constant source of inspiration, Nicole Ardoin, leads the Social

Ecology Lab at Stanford University. She and her colleagues have been working with the Girl Scouts of Northern California on energy conservation since 2013. The girls participate in an energy-saving education program that encourages them to think about and practice positive, small steps, like using cold water in washing machines.

Only the children participate in the program, and yet Nicole and her research partners confirm that what the girls learned positively impacted actual energy-saving decisions in their homes. According to Nicole, "We found that engaging people in reconsidering their energy use is a family affair. Fourth- and fifth-grade Girl Scouts shared their newfound interest, enthusiasm, knowledge and skills with their families. In turn, that information and interest diffused to parents."[3]

The program saves the families money, generates positive conversations between kids and their parents, and contributes significant energy savings. The researchers have done randomized control trials on this project that confirm that over time, this energy efficiency behavior persists.

Air-quality apps create momentum for China's sweeping clean-air reforms

Access to real-time data helps us, as individuals and families, to behave in climate-friendly ways. It also drives policy changes at astonishing scales. If you've been to China in recent years, you'll have seen how omnipresent cellphones have become. According to China Internet Watch, the country has more than 854 million internet users—more than the entire population of Europe. Many of these people begin their day by checking an air-quality app. With data available from tens of thousands of sites across four hundred cities on the mainland, these apps have become as much a part of daily life in China as its social media platforms. Air-quality apps enable people to make better-informed decisions about whether they need to stay indoors or put on a face mask, or if the air is clean enough for them to bike to school or work. These apps provide a vital source of information in a country notorious for deadly air pollution. They also supply powerful motivation to demand positive change.

Air pollution hit highs never before seen in China in 2013. In the capital city of Beijing, average particulate matter pollution (PM) was nine times the amount considered safe by the World Health Organization. In January of 2014, pollution reached thirty to forty-five times the recommended daily levels, and residents were told to stay indoors. Seventy percent of the entire Chinese population experienced PM levels that, if sustained, would correspond to a 6.5-year drop in life expectancy for the average person. The same year, a report by the Shanghai Academy of Social Sciences said that pollution in Beijing was so bad, the city was nearly *uninhabitable for human beings*.[4]

An influential study published in the *Proceedings of the National Academy of Sciences* found that air pollution had shortened the lifespans of people living in northern China by five years compared to those living in the south.[5] The study went viral on social media and was carried in major news outlets in China and internationally. Such dire health impacts, in combination with daily app reminders of how bad the situation was, fueled an outcry and drove mass public demand for change. In this case, personalized awareness of how critical the situation was led a national response.

In 2014, Premier Li Keqiang declared a "war against pollution" at the National People's Congress. The declaration came just a few months after the government instituted the toughest-ever clean-air policy in the country, allocating $270 billion to specific plans to reduce annual average PM. Beijing, which had already set aside $120 billion to fight pollution, would need to reduce PM levels by 34 percent to meet its targets. More than half of China's air pollution comes from coal-fired power stations. The national action plan on air pollution created an outright ban on new coal-burning plants, and accelerated the use of scrubbers and filters. It prohibited large construction projects in order to prevent smog from cement production and diesel trucks. It created a new environmental protection agency with tough, far-reaching powers of enforcement.[6]

China has achieved remarkably cleaner air—in just four years. Between 2013 and 2017, PM levels in Beijing dropped by 35 percent, while levels in surrounding regions dropped by 25 percent. According to a 2019 UN report, "No other city or region on the planet has achieved such a feat."[7]

As a result of this profound turn of events, the average Chinese citizen could increase their lifespan by 2.3 years relative to what it would have been in 2013, as long as these reductions in air pollution are sustained.[8] There is crucial work still to be done, especially beyond major cities, but China is committed to staying the course. The shift from *air-pocalypse* to poster child for combating air pollution has been fast and decisive. In November 2019, Beijing was removed from the list of the world's top two hundred most-polluted cities.[9]

Seeing positive change occur so quickly, and at the scale of a country the size of China, is deeply hopeful. It demonstrates that when scientific data, political will, public engagement through personal devices, and clean technologies work synergistically, remarkable achievements happen.

China's success with curbing air pollution is just one of many positive shifts happening at vast scales. Indeed, it's likely you're involved in another positive revolution, and it's as close to you as the food you eat today.

Bet you're eating more plants these days

You probably eat more plant-based foods than you did even a few years ago. For one thing, plant-based foods are way more accessible, and they come in so many more tasty choices. Whether you are hankering for fast food, craft beer, a fancy dinner, or something to pack in a school lunch, more vegetarian and vegan options are on restaurant menus and supermarket shelves. The surging demand for plant-based foods is being driven by those under-thirties I mentioned at the beginning of this chapter. Their food choices reflect their top-of-mind concern for animal welfare, the environment, and personal health.

Transforming how we eat is crucial to combating climate change and biodiversity loss. Research published in the journal *Science* in 2018 says that food production accounts for more than a quarter of all global greenhouse gas emissions.[10] If we include emissions caused by processing, transport, storage, cooling, and disposal of food, then the number rises to more than 40 percent, according to the World Economic Forum. How and where we produce food has the biggest impact of any human activity on the planet.

Some food industry analysts credit Gen Z's presence on Instagram with the rise in popularity of veganism. Widely shared images of beautiful "real food"—green goddess smoothies, cauliflower buffalo wings, Thai curry pumpkin soup, vegan apple crumble tart with salted caramel, and thousands of other tasty delights—convey plant-based eating as something colorful, delicious, healthy, good for the planet, and on trend.[11]

Gen Z are also using their personal devices to source foods that fit these values. In a 2019 analysis of tens of millions of online orders placed by more than 21 million diners using Grubhub, seven of the top ten orders were vegan and vegetarian dishes.[12]

In North America, 2019 was a breakthrough year for plant-based eating. Canada's official food guide downplayed animal fat and protein, promoting whole plant foods as the foundation of healthy eating. American media outlets, including *Food Business News* and *Forbes,* named "plant-based foods" the trend of the year. Sales of plant-based foods grew by more than 11 percent in the US in 2019, and analysts say this is just the beginning of a massive growth period as plant-based foods become even tastier and more consumers change their eating to match their desire for more sustainable options.[13]

Plant-based eating is rapidly becoming a global trend too. Analysts report declines in meat consumption in the UK and across many countries in the European Union, as well as significant rises in people choosing vegetarian and vegan options.[14] Meanwhile, plant-based diets are a leading health and wellness trend across Indonesia, India, and other Asian countries.[15] According to the New Zealand Institute for Plant & Food Research, nearly 40 percent of consumers in China want to purchase foods that are healthy for themselves and for the environment, and are reducing their meat intake in favor of tofu, vegetables, and vegan meats.[16]

The meteoric rise of plant-based eating is a super hopeful trend for the planet. It's a big step forward in achieving the recommendations of the world's first full scientific review of what constitutes a healthy diet and sustainable food production. In 2019, the EAT-Lancet Commission on Food, Planet, Health brought together thirty-seven world-leading scientists to answer the question: "Can we feed a future population of ten billion people a healthy diet within planetary boundaries?" The answer is yes. According to the report, by changing the way we produce, transport, and consume food, and reducing food waste, we could feed everyone a healthy diet while improving the health of the planet. By switching to plant-based diets, we'll lower the risk of cancer, stroke, and diabetes, and in so doing, avoid eleven million adult deaths per year.[17] The contribution of plants is so vital to our health, to feeding global society, and to the planet, the United Nations dedicated 2020 as the International Year of Plant Health.

While plant-based food is on the rise, the global demand for meat is also increasing, particularly in developing countries where people are better able to afford meat as they grow richer. That's a big concern because, as the World Resources Institute puts it: if cattle were a nation, they'd be the world's third-largest emitter of greenhouse gases, after China and the us.[18]

Yet rising meat consumption may not be a foregone conclusion. Plant-based or "meatless" meat, like the Beyond Burger and the Impossible Burger, is a fast-rising trend. KFC reported sales of more than one million vegan burgers in the first month they were on sale.[19] Burgers, chicken nuggets, and other foods made from plants that are meant to taste like meat are in line with the priority young people all over the globe are placing on the environment. The "meatless" meat industry is projected to grow to $140 billion in a decade.[20]

Thanks to delicious plant-based options increasingly filling the food categories once reserved for meat, and the popularity of flexible approaches like Meatless Mondays or Vegan Before 6:00 or Veganuary, more and more people now identify as "flexitarians"—folks who mostly eat a plant-based diet but enjoy meat on occasion, and "pescatarians"—semi-vegetarians who abstain from eating all animal flesh except for fish. In 2019, more than a third of American households had at least one family member who followed a vegan, vegetarian, pescatarian, or flexitarian diet. The percentage was even higher for Gen Z and millennial households.[21]

Personal devices are helping to take a bite out of food waste

One of the great opportunities for tackling climate change and improving the health of people all over the planet is stopping food waste. Globally, nearly one third of food is lost or wasted. According to the Food and Agriculture Organization of the United Nations, 45 percent of all fruit and vegetables, 35 percent of seafood, 30 percent of cereals, and 20 percent of both meat and dairy products—1.3 billion tons of food per year—is wasted in a world where more than 10 percent of people struggle with hunger. It also carries a huge environmental and economic cost, because producing all that unused food demands resources and contributes to soil erosion, deforestation, water and air pollution, and greenhouse gas emissions. Plus, food that's thrown in the garbage ends up in the landfill, which creates emissions too. But in the age of personalization, tech innovations like smart kitchens, sensor devices that detect food spoilage, and tracking systems are having a big impact.[22]

Fridge Pal, Foodfully, and similar apps help you make shopping lists, track expiry dates, and suggest recipes based on the food you have at hand. Companies like Leanpath do the same thing for commercial kitchens, using food-waste smart meters to track and identify areas of overproduction. The hospital at the University of California, San Francisco, cut their food waste in half, and now directs unused food to charities in the city, thanks to a combination of this high-tech tracking system and personalized menus that allow patients to choose their own meals.

Often food is wasted before it even makes it to its final destination—but again, personalized technology is stepping in to solve the problem by making it easier to redistribute unwanted food to those who need it. Food Cowboy matches transport trucks with food they need to unload—say, pallets of overripe tomatoes—with charities happy for a donation. The food gets eaten, and the shipper gets a donation credit.

With Transfernation, an on-demand food redistribution service operating in New York, food left over from receptions, weddings, and business meetings is picked up by volunteers, or by transportation networks like Lyft or Uber, and delivered to food banks and shelters. Donors get a tax receipt, deliverers get fifteen dollars a trip, and people in need get delicious food.

Healthy produce often gets thrown away for cosmetic reasons before it even reaches the grocery store. Instagram images of ugly yet adorable-looking fruits and vegetables sparked the @UglyFruitAndVeg campaign to encourage people to see less-than-perfect produce in a more accepting light. Imperfect Foods works directly with farmers and retailers to source ugly produce that would have been thrown away, and deliver these healthy items to your doorstep for up to 30 percent less than buying food in the grocery store.

In cities all over the world, you can download apps like Feedback, goMkt, Food for All, and Too Good To Go that alert you to restaurants nearby offering deep discounts on meals for pickup near closing time. NoFoodWasted in the Netherlands does the same thing, but for grocery shopping, alerting you to supermarkets offering discount prices on products reaching their best-before dates. And if you're moving to a new house or going on vacation, or you've simply got food in your fridge you know you can't use in time, you can donate it to your neighbors—or ask for that egg you need for a cake—using Olio. This app promotes free food sharing, and its more than 1.5 million users have already shared nearly three million portions of food since it launched in 2015. By making things accessible in real time through our personal devices, these innovations are actively reducing greenhouse gas emissions, building more equitable systems, and deepening our sense of belonging to communities that care about making positive changes for the planet.

Personalized technologies support collective campaigns

In the Age of Personalization, we participate in environmental action in ways that reflect our diverse identities and shared passions. Climate change activism has become a cultural movement. Pink Floyd's David Gilmour raised $21.5 million to fight climate change by selling his guitar collection. Lil Dicky and dozens of famous musicians released "Earth," using their music and their huge followings to gain vital coverage of climate change. Coldplay chose to end tours until they can be environmentally sustainable, while Massive Attack has commissioned the Tyndall Centre for Climate Change Research to provide guidance on how to decarbonize music tours, which they will share with others.

When we get individualized feedback on our personal devices, enhanced by access to real-time analysis, it helps us understand how our everyday actions impact the planet. We also have many more ways to connect with others to make meaningful changes. If you've used a footprint calculator to measure your ecological impact, for instance, you know how bad airplane travel is for the environment. A round-trip flight between New York City and London, according to the German nonprofit atmosfair, generates more CO_2 per passenger than the average person in one of fifty-six countries emits in a whole year.[23] A rapidly growing number of people now cite climate change as the reason they are switching from flying to taking trains (night trains in Europe are making a comeback) or refusing to travel at all if other options aren't available. Flight shaming has become a real force for change. JetBlue has committed to going completely carbon neutral in 2020 on all flights in the US. In addition to using a more sustainable fuel made 100 percent from waste and residue raw materials on some of their flights, they'll achieve carbon neutrality by investing in carbon offsets—donating money to environmental projects including forest conservation and solar- and wind-power farms.[24] Harbour Air, North America's largest seaplane airline, is transforming its fleet into an all-electric commercial fleet. EasyJet is now offsetting emissions from all of their flights. Meanwhile a technological race is on to make solar-powered plane travel and other eco-friendly flying options feasible.

Köpskam?

Fashion trends come and go but here's one that might surprise you. The newest thing to hit your closet is, well... not new! Secondhand is projected to be a one-and-a-half times *larger* business than fast fashion in ten years, according to a 2019 report from the online clothing store thredUP and retail analytics from Global Data. Used fashion will skyrocket in value to $64 billion in the US, the report predicts, while the market for new fast fashion from brands like H&M and Zara will only reach $44 billion.[25] Buying resale appeals to the 74 percent of eighteen-to-twenty-nine-year-olds who prioritize sustainability when shopping. Buying one used item reduces its carbon footprint by 82 percent.[26] By 2028, the proportion of used clothing in your closet will have doubled.

The massive shift toward the reuse of our clothing decreases waste, and prioritizes durability over disposability and planned obsolescence. It's a real positive step forward in tackling climate change. According to the United Nations, the fashion industry currently produces more carbon emissions than all international flights and sea shipments put together.[27] That's a whopping 10 percent of global greenhouse gases, plus it contributes 20 percent of the world's wastewater.[28] And 35 percent of microplastics released into the world's oceans are from the polyester, acrylic, and other synthetic textiles that make up nearly two-thirds of our clothes.[29] Shopping for new clothes is so destructive to the planet, the Swedes have a new term to keep you from doing it: *köpskam*—the shame associated with shopping or consuming.[30]

Keeping up with the Joneses—but for good

Anyone with a social media account knows we measure ourselves against what others are doing, which explains why people are more likely to install solar panels in neighborhoods where they already exist. In a study involving 1.4 million residents across the US, people who had solar panels themselves were 62.8 percent more likely to convince others to get them.[31]

In our hyperconnected world, we see what our friends and other people we relate to are doing. This is a powerful lever for environmental change. The beliefs of the groups we see ourselves as a part of have more influence on what we do than the words of a scientific expert, no matter how well informed, according to Dan Kahan of Yale University. He uses cultural cognition theory to help us move beyond thinking that evidence-based knowledge is all that matters to communicating about the climate crisis or biodiversity loss. Clearly that information is extremely important, but it isn't enough. We conform our beliefs to the values that define our cultural identities.

That's why the increasing diversity of voices we now hear speaking out for climate change action is so important. The combined actions of diverse groups of ordinary people are driving the mass climate change movement. We *act* when we see people we relate to also taking action.

Our increased capacity to use personalized technologies to engage with the environment in whatever combination of face-to-face and virtual communities we relate to enables diversity to flourish. People brought together by shared interests, faiths, or identities are making eco-consciousness part of their community values. These community-building collectives include Red, Bike and Green, which creates meetups for black urban cyclists across the us. Green Muslims grew out of a zero-waste potluck iftar during the month of Ramadan. Greener Pride, an initiative of out for Sustainability, works with organizers to move toward carbon-neutral, zero-waste Pride celebrations across the globe. Interfaith Power & Light, an influential multifaith organization, draws together individuals of different faiths who have a common purpose in mobilizing against climate change. As Dan Kahan puts it, "What we need isn't more evidence, but people seeing other people who they identify with *acting* on the basis of the evidence." [32]

IN OUR PANIC over the state of the planet, it can feel as if everyone needs to change everything. Whenever I hear a well-meaning adult ask a group of kids, *What do you think we should do about climate change?* I can't help but think how misguided that question is. If we were in a surgical theater, would we ask a kid, *What do you think we should do about this brain tumor?* Inclusion and abdicating responsibility are very different things. Rather than asking each of us to be or do all things, the Age of Personalization enhances our capacity to target and tailor our contributions within broader networks of people with other specialized skills and routes of influence. A sort of communal intelligence emerges when we use our devices and/or connections with other people to engage with these issues—personalization enables us to have a collective effect that would be impossible as individuals. Hope exists in delivering strategic responses together.

6

WE ARE NOT
THE ONLY ONES ACTIVELY
RESPONDING

Wolves make rivers
Salmon grow forests
Seabirds fertilize coral reefs

Every last one of us is
made from new stars
and old stars

We are always
becoming.

I N 2019, RESEARCHERS at Tel Aviv University recorded ultrasonic squeals made by tomato and tobacco plants stressed by lack of water or by having their stems cut. In recent years scientists have revealed the astonishing capabilities of plants to see through photoreceptors in their leaves and stems; to taste through their roots; and to smell, not through noses, but through their genes, via receptors to volatile chemicals. Now, for the first time, plants have been recorded making airborne sounds.[1]

I keep an ever-growing file of scientific discoveries, like these, that demonstrate the astonishing capacities of other species: archerfish can tell one human face from another; manta rays recognize themselves in the mirror; rats giggle when they are tickled; bats eavesdrop on feasting neighbors to find food; chimps and orangutans experience midlife crises; frigate birds sleep while flying; North American brown bears make a paste of osha roots and saliva to rub through their fur to repel insects; mother trees preferentially direct nutrients to their relatives; humpback whale babies "whisper" to their moms to prevent predators from hearing them.

Birds carry no provisions, yet they survive winter storms. Eastern gray whales have no hands with which to hold their babies as they cross busy shipping lanes, yet increasing numbers of them successfully navigate one of the world's longest migrations, from Mexico to the Arctic, each year. Sea otters restore kelp forests, and eelgrass beds as well. Animals and plants are not the helpless victims we so often portray them to be. This doesn't remove any of our responsibility to care for and repair the Earth, but it should change the way we consider the needs and extraordinary contributions of other species.

Yet we're often so focused on humans, we fail to notice other species, including the plants that give us oxygen regardless of whether we're behaving well or badly. Botanists have coined the term "plant blindness" to describe our widespread failure to see the plants around us, despite our utter reliance on them. It's a bias we learn as we grow up in a human-centered, and to some degree animal-centered, world. If you want to test this for yourself, pull up a photo on the internet of an elephant standing in a natural habitat of grasses or trees. Show it to a friend and ask them, "What's this a picture of?" Dollars to donuts most of them will say, "An elephant." They will not mention the trees or grasses, let alone name them, because ever since they were little kids, people have been showing them photos like this and pointing out the elephant, and rarely, if ever, the plants.

The current scientifically accepted estimate is that there are 8.7 million other species on Earth. But the truth is, no one knows for sure. Life is far too wonderfully diverse, complex, vast, and, in many cases, miniscule in size to tally. Scientists say estimating the number of species on the planet is one of the greatest challenges in biology. Current estimates range somewhere between 5.3 million[2] and, according to a 2016 study, one trillion species.[3] If this upper number turns out to be correct, it would mean that only 0.0001 percent of all life on Earth has been identified. Only 1.5 million species on Earth have been scientifically named.

You are nature

When we dare to look beyond our human-centered view, it's obvious our lives are interconnected with other species in myriad ways. We drink the same water dinosaurs drank. Every time we exhale, we spread pollen that might grow to become a plant. Where is the line between human and nature when the bacteria living on or in our bodies equals or outnumbers our human cells?[4] The bacteria that inhabit "planet you" exist as organized communities, passed down from your grandmother and onward to your children, evolving with you. These microscopic life-forms are powerful enough to affect how your body functions at the most fundamental levels. They, in turn, must adapt to every change in your diet, health, and lifestyle. The microbial communities living on and in Earth's nearly 7.8 billion people drive and are being driven by human evolution.

We think of the history of agriculture as one of human dominance over other species. In *The Botany of Desire: A Plant's-Eye View of the World*, Michael Pollan turns that notion on its head, driving home the message that we exist within a great reciprocal web, by showing how apples, tulips, and even cannabis have caused us to go to sometimes extraordinary lengths to help them grow and reproduce.

The animals we welcome into our homes also exert influences we may not recognize. When a cat purrs, it triggers an inherent mammalian sensitivity in humans to nurture offspring. You feed the cat because the cat signals you to do so. You in turn benefit from the relationship. Cat ownership has been shown to cut your risk of stroke and heart disease by as much as a third.[5] And when you gaze into your dog's eyes, you feel a rush of love because the mutual gazing raises oxytocin levels. In 2015, a team of Japanese researchers reported that not only do you experience this positive rush of oxytocin, but the dog does too, which may explain the coevolution of the human-dog bond.[6]

The lives of other species entangle with our own. Like us, their livelihoods shape and are shaped by political, economic, and social forces. Coyotes learn to look both ways before crossing streets. They use sidewalks and cross in pairs, according to Stan Gehrt, a wildlife ecologist at Ohio State University who runs the Urban Coyote Research Project. Contrary to popular belief, they feed primarily on small rodents, rabbits, deer, and overwintered fruits like crabapples, and not on small pets or garbage. Coyotes have successfully moved into every large metropolitan area in North America, and the vast majority of them coexist with people in densely populated cities without conflict.[7]

Bald eagles, too, are on the rise in urban settings. Their numbers along the coast of North America's Pacific Northwest region are on the rise, not solely due to targeted conservation plans and the banning of DDT, but in part because the birds are hunting seagulls who are attracted to the landfills associated with this rapidly urbanizing landscape.

Recognizing animals and plants as active agents challenges the tacit assumption of human exceptionalism. A plethora of animal species are now known to make and use tools; to innovate problem-solving approaches; to improvise and learn from one another; and to experience emotions that range from joy and happiness to sadness, grief, jealousy, and the desire for revenge. Humans are not the only species to have unique personalities or lead purposeful lives. Dolphins, elephants, and great apes think, know, and remember in ways so close to our own that they've been dubbed our "cognitive cousins." They recognize themselves as individuals when they look in a mirror. They mourn the loss of friends and relatives who die. Populations of killer whales in different parts of the world have distinct languages, social structures, and hunting approaches. Animal cultures—and cultural variation across highly intelligent, social species—abound.

We are not the only ones actively responding to crisis. And we aren't the only ones whose lives are shaped by networks of relationships. Satellite technologies enable wildlife biologists to track families of sperm whales, populations of elephants, and colonies of gannets, contributing to an avalanche of recent research that highlights the complexity of social relationships and cultures in animals—and plants. Trees communicate with each other through fungal (or mycorrhizal) networks between their roots. Trees also provide habitat for other plants and animals. When we leave enough trees to support these networks of other species—and mycorrhizal networks—forests can regenerate and thrive.

Social networks between trees promote faster recovery of forests. Monkeys catch on faster to new foraging techniques when they have stronger social networks. And networking among humpback whales can explain how many populations have returned to near historic numbers, as we saw in chapter four. In fact, humpbacks recovered faster after the end of commercial whaling than researchers forecasted, despite contending with all kinds of issues, including shipping lanes, plastic pollution, and increasing underwater noise levels that have doubled in intensity every decade since the 1950s. Fred Sharpe, a humpback whale researcher, says the reason humpbacks recovered so quickly is that they are great social networkers. If one group of humpbacks is using a hunting technique that is more effective, other humpbacks have a "grass is always greener" attitude that makes them try it out for themselves. They imitate and use successful innovations to their benefit.

Inheriting resilience from mom and dad

Coral reefs are highly threatened by climate change. In recent years we've witnessed wide-scale coral-bleaching events, with mass die-offs of corals and impacts on the fish who live within these communities.

Yet remarkably, research reveals that some baby coral reef fish inherit tolerance to warming oceans, with help from their moms and dads. "What we are finding," says lead researcher Philip Munday (professor of coral reef ecology at James Cook University in Australia), "is a surprising amount of potential for fish to cope." This example is so complicated and representative of the unexpected capacities of other species, I'd like to spend a bit of time explaining it in more detail.

Most coral reef fish are negatively affected by a water temperature increase of just a degree or two above the summer average. The change can impact their swimming ability and growth rate, and—since warm water holds less oxygen—their capacity to breathe. Yet, when Philip's lab reared spiny chromis damselfish parents, and then their offspring, in water that was three degrees Celsius (5.4°F) warmer than current-day ocean temperatures, the offspring had a perfectly normal capacity to get enough oxygen. The baby coral reef fish were able to handle higher water temperatures when their parents were previously exposed to those conditions.

It turns out the capacity for offspring to benefit from their parents' experiences doesn't just happen with fish. *Daphnia*, often called a water flea, is found in freshwater lakes, ponds, and puddles. The tiny crustacean can hatch with either a round head or a pointed head. If it shares the water with predators such as fish, midges, or other insects, spikes and spines help lessen its likelihood of being eaten. For many species of juvenile water fleas, whether or not they grow a defensive helmet depends on the experiences of their mother. "If a mother *Daphnia* has been exposed to chemical cues of a particular predator while she is pregnant," Philip explains, "then she produces lots more offspring with pointy heads."

Such astonishing malleability exists in other species too. "If aphids are in a habitat with predators, their babies can develop wings and fly away. They have the plasticity to have wings or not," says Philip.

It sounds like a fairy story: the environment changes and the hero grows wings! But the dark force driving Philip Munday's research is all too real. Changes in climate are already exposing fish to warmer ocean temperatures and ocean acidification. What he wants to know is how animals are affected, and if they're able to adapt. Part of the answer, Philip says, lies within the science of epigenetics.

Epigenetics is the study of changes in gene *expression*. This differs from a mutation in that the actual sequence of a gene— the underlying DNA code—remains unchanged. Think back to those damselfish. Placing the parents in warmer water doesn't change the basic genetic code they pass on to their babies. But it *can* influence which genes are expressed. It's the exposure of the parents to the higher water temperatures that appears to determine which genes in the DNA sequence are turned on or off. The parents already had to have the capacity to deal with warmer temperatures somewhere within their existing DNA in order for it to be turned on, or expressed, within their babies.

The potential for differences in gene expression depends upon the range of forms that already exist within embryos in a wild population. And that's what makes coral reef fish such great study subjects for epigenetics—they embody some of the world's most amazing examples of plasticity. Bluehead wrasses, for example, can change gender. Females can become large, colorful, territorial males. Other bluehead wrasses exist as "sneaker" males that look identical to the females. This camouflage strategy enables them to sneak in undetected and spawn with the females within the territories of larger males.

The gender a young bluehead wrasse will manifest depends upon the number of individuals it encounters when it is first developing. If it hatches on a large reef where there are so many fish that it's difficult for the big males to defend harems of females, then the developing baby is more likely to become a small sneaker male. "Social interactions really early in life set off these trajectories of whether they become a female, whether they become a male, or whether they become a female that then changes sex to become a big male," says Philip. "This one species has all these morphologies, all these alternative strategies, all this plasticity—it's just incredible."

Yet just because one sees advantageous plasticity in one trait, it doesn't mean it will exist in all traits. In Philip's study, the spiny chromis damselfish offspring were perfectly capable of getting enough oxygen in those higher water temperatures, but they did not breed.

Here is where the hopeful story of damselfish resilience to warming ocean temperatures suddenly veers toward despair. Or does it? Thanks to a smart-thinking postdoctoral fellow in Philip's lab, the experiment included an additional treatment option. Rather than just putting the fish directly into water that is three degrees Celsius warmer than normal, Jennifer Donelson put them into water 1.5 degrees Celsius (2.7°F) warmer for one year and then into water three degrees Celsius warmer the second year.

And those fish did breed.

A gradual rate of temperature change enabled the damselfish to acclimate in ways they couldn't do in response to a sudden increase. In other words, given enough time to adjust, the kids were all right.

Clearly, this doesn't mean we shouldn't try to stop the ocean from warming. Thirty percent of corals in the Great

Barrier Reef died in a horrific nine-month marine heat wave in 2016. An increase of just one to two degrees causes corals to bleach, and if they stay bleached, they will die. In 2016 and 2017, the Great Barrier Reef experienced its worst-ever back to back bleaching events. We can recognize the urgency of the problem *and* be inspired by the resilience of other species.

Do corals themselves have the potential to gain resilience from their parents' experiences? In a series of experiments at Hollie Putnam's lab at the University of Rhode Island, researchers exposed adult corals to increased temperature and acidification, and then exposed their offspring to the same situation. "We found that there is potential for beneficial acclimatization," Hollie says. "The offspring show greater survivorship and growth rates if the parents have been previously exposed to short periods of adverse conditions." Hollie believes that here, too, the rapid adaptation is due to epigenetics.

Are the results too good to last?

But how long will these benefits last? Is the improved capacity to live in warmer water a fleeting phenomenon, like Cinderella's pumpkin coach, or is it passed on through future generations?

Thousands of miles north, and on the other side of the globe, scientists in Maritime Canada studying winter skates—the fish, not the hockey equipment—found an intriguing way to explore that question. They studied genetic material from two populations of these cartilaginous fish—one that lives off the exposed coast of Nova Scotia in the chilly Atlantic waters, and one that lives in waters ten degrees Celsius (18°F) warmer in the southern Gulf of Saint Lawrence. The winter skates in the Saint Lawrence are dramatically smaller, an adaptation that enables them to live in the lower-oxygen conditions of the warmer water. Their small size was achieved by epigenetically modifying their genes to make a physiological change.

These two groups have been geographically separated for seven thousand years, yet they are genetically indistinguishable. Technically, the smaller skates have not evolved—their DNA is identical to the other population of skates. Instead, they've adapted their bodies to the warmer water by making a total of 3,653 changes in their gene-expression patterns. These epigenetic changes have developed and persisted over the course of 318 generations.

And yet, because they do not involve permanent changes to the genetic DNA code, epigenetic changes are also potentially reversible—a quality that is driving a revolution in human medicine. "The great potential for epigenetic therapies," write Nita Ahuja and her colleagues from Johns Hopkins University School of Medicine in a 2016 paper in the *Annual Review of Medicine*, "lies in the fact that, unlike genetic abnormalities, epigenetic changes are reversible, allowing recovery of function for affected genes with normal DNA sequences." In other words, epigenetic therapies help reprogram cancer cells to return to the straight and narrow.

Today a vast collection of medical and biological studies demonstrates how changes in temperature, along with hormones or other chemicals, affect DNA by modifying certain epigenetic factors, which instruct the genes whether or not to be expressed. It's a far cry from what people understood over a century ago about how environments create change. Back in the 1880s, German biologist August Weismann sought to understand how changes to one generation might affect the next by amputating the tails of five successive generations of mice. He concluded that the effects of an environmental stimulus (in this case, cutting off a tail) could not be transmitted to its offspring (the mice were all born with tails).

What Weismann failed to realize is that cutting off a parent's tail has nothing to do with whether or not a mouse will be born without a tail. Temperature or chemicals, not physical injury, trigger epigenetic factors. Plus, the capacity to be born without tails would have had to already exist within the genetic code of mice in order for this trait to ever be passed along or expressed.

That said, astonishing examples of transgenerational resilience that defy belief continue to be discovered. For example, Mylene Mariette, a behavioral ecologist at Deakin University in Australia, was observing relationships between zebra finch parents when she heard a call she didn't recognize. Although she did not realize it at the time, it was the sound of a zebra finch calling to its chicks—chicks that hadn't hatched yet. Zebra finch parents were communicating to their babies still inside the egg.

"At first I didn't know the call had anything to do with temperature," Mylene explains. "When you discover something really new, you don't have a theory to go with it. It was quite an adventure." Yet after listening to six hundred hours of recordings, she determined that zebra finch parents only make this call when it's super hot outside and they're panting to cope with the heat. "The parents mostly make the hot calls near the end of the incubation period," she says. "We don't know why, but perhaps the embryos have not developed the ability to hear before then."

In the experiments that followed, Mylene and her research colleagues placed eggs in incubators and divided them into two groups. One group heard typical parent-to-parent calls and the other was played recordings of parents making "hot calls" for the last five days before hatching.

The chicks showed no difference in body size at hatching. Yet within twenty-four hours of the chicks being placed in either a cool or hot nest, the research team could already detect a difference in their weights. And it continued throughout their development. Chicks that had heard the hot call as embryos grew more slowly and had smaller body sizes. Smaller bodies are easier to cool, which is an advantage in hot temperatures. When they grew up, the hot-call birds went on to prefer warmer nesting spots. And they had more babies than the birds that hadn't been primed for the heat.

In short, the research indicates that chicks that hear hot calls during the time their temperature-regulation system is forming develop differently. It's the first time such profound effects of prenatal acoustic experience have been documented. Zebra finches appear to be using particular songs that alter the unhatched chicks' temperature-regulation systems and change their sensitivity to heat over the long term. They seem to prepare their unhatched chicks to deal with high temperature extremes—the kinds of temperatures we expect from climate change.

"My research makes me hopeful," Mylene says, "because it really shows that birds are better at dealing with changes to their environment than we think. Perhaps they will be able to deal with increasing temperatures and all those other environmental changes that are happening too quickly."

I AM AWED by the capacity of damselfish and corals and zebra finch parents to positively impact the resilience of their off-spring in these daunting circumstances. It fuels my already fervent passion to reduce carbon dioxide emissions.

Recognizing transgenerational resilience is inextricably linked with hope. It offers possibility, even in the midst of extreme droughts and rising ocean temperatures. I share this hopeful perspective with my friend Lillian Howard, who is from the Mowachaht/Muchalaht First Nation and of Nuu-chah-nulth, Kwakwaka'wakw, and Tlingit ancestry. She is a strong advocate for Indigenous, social, and environmental justice. I tell her how much it matters to me to know that other species gain resilience from the experiences of their parents. She responds with a warm smile.

"Everything is interconnected," Lillian reminds me. "Our ancestors are always with us."

7

THE STRENGTH OF EMPATHY, KINDNESS, AND COMPASSION

A hummingbird
alights
at the very top
of the snow-drenched cedar.
Not a drop of nectar in this winter snowscape.

Birds have no pockets,
no provisions.

How would it feel
to embody such courage?

I N RESPONSE TO the massive threats facing the planet, increasing numbers of people are engaging in strategies to create a more just, sustainable, and ecologically diverse world. Millions of organizations have sprung up in the past three decades to do just that, collectively seeking to advance rights and opportunities; advocate for legal protection; invent and restructure institutions; and conserve and protect species and ecosystems. In his beautiful book *Blessed Unrest*, Paul Hawken describes the sheer number of organizations and individuals pursuing social change that have emerged in the past thirty years as "the largest movement the world has ever seen."

When we act on behalf of the common good, we feel proud, and that pride is important to environmental engagement. Over the past few decades there has been a dramatic increase in the number of research studies exploring the role of emotions in behavior change. And what those studies reveal is that feelings of pride, or compassion, or that you are part of something meaningful, are much more likely to keep you engaged in changing your behavior than if someone is telling you there is no hope. Princeton University research demonstrates that emphasizing the pride people will feel if they make environmentally conscious decisions is a better way to promote eco-friendly behavior than making people feel guilty for not living more sustainably.[1]

You are hopepunk

When you stand up for other people and other species, you are adding your voice and actions to a vast movement focused on making change for the better. Hopepunk is a narrative of positive resistance. We see evidence of hopepunk in families and communities that welcome and support refugees. We see it in Finland, where people do not sleep on the streets. A dramatic change in policy recognizes "home" as a basic human right and now gives people housing as soon as they need it. Hopepunk fuels the protest marches against climate change, racism, inequality, and human-rights injustices. We see it in mission-driven institutions, like the Monterey Bay Aquarium, which successfully mobilizes big-scale social change around marine protected areas, sustainable seafood, climate change, ocean plastic, and more by working in diverse partnerships, and modeling ocean conservation values in everything they do. Hopepunk shines through in the rising tide of people who volunteer, and in those who help friends and neighbors. Americans, for instance, volunteered a record $203.4 billion worth of time in 2018.[2]

The term "hopepunk" emerged within pop culture in a Tumblr post by Alexandra Rowland, a Massachusetts writer, in July 2017.[3] She positioned it as an antidote to "grimdark," which includes the apocalyptic genres we know so well from films and video games. And of course, there is also "noblebright," which is basically the storyline of fairytales—a powerful hero comes along and saves us. I think a lot of people have positioned Greta Thunberg in a kind of noblebright narrative.

Rather than waiting for a single heroic figure to lead us out of trouble, hopepunk situates heroism as a collective response. It's about committing to what you believe in and acting as a force for good. Regardless of how wrecked something is or how bad things might be, you act in the best ways you can. Hopepunk acknowledges that caring about something requires strength and bravery.

Vulnerable, inspired, defiant, responsive, connected to others present and elsewhere—these are just some of the feelings Jason Miller, a graduate student, described as he shifted from being a sympathetic bystander, sending support through the internet, to standing in solidarity with protesters demanding human rights. "The experience finally gave me a bodily response to what I have known intellectually for some time— that this goal of affecting positive change will always be an uphill battle, and that one must be willing to play the role of the outsider even when advocating for a collective. But I also left feeling the vitality of those desiring change."

Dystopias thrive in an unjust world

Our feelings about the state of the planet are intimately tied to our conceptions of utopia and dystopia. The narrative of planetary doom and gloom assumes a universal decline into a dystopian future. Yet the vast range of issues changemakers seek to address reminds us that dystopias are not only a future threat but a current reality in places all over the world. Speaking on CBC Radio in February of 2020, Dr. Mohamad Abrash, a surgeon in Syria, described the horror experienced by the more than 700,000 civilians trapped by a closed Turkish border and below-freezing temperatures while trying to flee the fighting.[4] "Nowadays we have lost the future," he said. "We are living day by day."

Such terrible situations occur because of systemic inequity and injustice, which too often inflames political, ethnic, or religious tensions. We live in a world where power and wealth are highly unequally distributed. As in the case of war, climate change creates displacement and enormous inequality. A range of studies show the global mismatch between the burden of climate change and those countries that are the highest greenhouse gas emitters.[5] Those living in poorer countries, who emit the least, will likely suffer the greatest impacts. This is inherently unfair, and unfairness is something people of all ages rail against, according to an influential 2017 study.[6] The millions striking for climate action make it clear that their demand is for climate justice. All over the world, we are witnessing large-scale mobilizations challenging old, unfair truths and demanding social transformation.[7]

Nurturing empathy and mindfulness to fuel hopeful environmental change

Jamil Zaki, director of the Stanford Social Neuroscience Laboratory, says empathy—the ability to both share and understand what other people feel—is not a fixed state but a disposition that can be learned. It can be cultivated, or it can wither through social isolation. Developing empathy is good for you, and for others.[8] People who have high empathy tend to be less depressed and lonely and better able to navigate tough social situations. If you are the partner of an empathetic person, you are likely to be happier in your relationship. Empathetic people tend to be kinder toward strangers and more likely to donate to charity or volunteer, which benefits the wider community. And, people with high empathy have higher engagement with environment and sustainability issues. Jamil argues that our rapidly urbanizing planet, and increasingly online presence, is causing more of us than ever before to lead more solitary lives. This lack of social connection works against the building of empathy and fuels the growing problem of loneliness.

Some have responded to this issue by looking at ways design can help promote trust and a feeling of belonging. The designers and researchers working for Happy City bring together the science of well-being with good urban design to help people act in ways that benefit the greater good—what psychologists call "prosocial behaviors." An experiment in Vancouver, for instance, shows that adding natural landscapes and color into public spaces increased people's feelings that strangers were more trustworthy. It also increased people's desire to care for the area.[9]

Other people focus on creating more prosocial mindsets. Over the past twenty years, training in mindfulness—the intentional cultivation of moment-by-moment nonjudgmental attentiveness to the present—has sparked what many are calling a "mindful revolution," in which meditation has joined healthy eating and exercise as the third pillar in wellness. Indeed, Apple named mental wellness and mindfulness apps as the number one app trend of 2018.

Climate change, biodiversity loss, and other challenges we face are too complex to be solved by technology fixes or governance alone. They require broader, cultural transformations toward sustainable ways of living, and that's where mindfulness comes into play.

Recent studies from neuroscience, psychiatry, psychology, and education suggest that mindfulness may support a fundamental shift in the ways we think about and act in response to economic, social, and ecological crises. Individual mindfulness coincides with compassion for others and the environment and a higher motivation to take climate adaptation actions that benefit the common good. Other studies show the positive impact of mindfulness on psychological resilience—our ability to rebound after adversity. Mindfulness has been linked to hopeful attitudes that help to steer us away from fatalistic defeatism and toward active engagement with environmental issues.

On a personal basis, an increasing number of studies show the influence of mindfulness on our feelings of connection with nature; compassion for the environment; pro-environmental values, intentions, and actions; and choosing to live more sustainably. Plus, the attention and awareness facets of mindfulness have been shown to sensitize people to the consequences of climate change and to increase their support for climate-positive policies and behaviors.[10]

Cooperation for the common good

The planetary crises we face are too vast to solve on our own. Luckily, the desire to cooperate often appears to be present, even in difficult situations, and we're naturally good at it, according to a 2018 focus issue of *Nature Human Behaviour*.[11] So how can we nurture cooperation for the common good? Part of the answer lies in helping people to see the impact of their own contributions to a greater whole, and for institutions to position cooperation as the more attractive option.

I saw the impact of using a cooperative strategy first-hand some years ago when I was working with a network of governmental and nongovernmental organizations across the us/Canada border involved in the Pacific Estuary Conservation Program (PECP). At that time, these conservation agencies found themselves competing for credit in order to satisfy their funders and members of the unique and key role they were individually playing. Infighting threatened to destroy the important collective work they were accomplishing. With the help of Ron Erickson, then director of the Nature Trust of British Columbia, I applied for a prestigious international award—the Ramsar Wetland Conservation Award—in the "effective partnership" category. The PECP won, and the acknowledgment of the network's collective impact was enough to carry the cooperative work forward for several more years.

The potential for post-traumatic growth

Just as the forests of Chernobyl embody both the wounds from their terrible history and the vibrance of new growth, the capacity to experience not only negative but also positive changes in the wake of crisis exists for us too.[12] Post-traumatic growth doesn't occur as a direct result of trauma. It's the personal struggle in the aftermath of devastating loss or crisis that is crucial to whether or not people perceive growth arising out of tragedy. People speak of having an increased appreciation for life and what they still have, and more clarity about their priorities and what is important to them. Some describe developing closer, more intimate and meaningful relationships. There can be a deep sense of personal strength *and* vulnerability—the lived knowledge that terrible things can happen and the discovery that one is capable of handling almost anything. Post-traumatic growth can occur, even as pain, grief, and distress endure.

Research tells us that most trauma survivors are simply trying to keep living, or questioning whether survival is worthwhile. They are not consciously trying to search for meaning or some sense of good that could come from the terrible things they experienced, which is why post-traumatic growth tends to surprise people when it occurs. Researchers say it is closely connected to the development of general wisdom and the way we modify the narrative of our life after trauma.

Judy Long, my dear friend and a palliative care chaplain, helps patients and family caregivers recognize and name losses, and "re-story" chronic sorrow. Gathering participants together in face-to-face and online classes, she guides family caregivers on how to find balance when caring for a loved one who is living with chronic serious illness. She also hosts workshops for physicians and other clinicians on professional grief and ways to meet difficult emotions and develop a positive growth mindset.

The importance of self-care
and caring for each other

We recognize the importance of self-care for firefighters and other rescue workers whose jobs put them in physically, mentally, and emotionally demanding situations. Yet rarely do we think in the same way of researchers who work on climate change or biodiversity loss; or teachers, parents, and kids who are grieving the state of the world; or the activists involved in climate justice. They too are at high risk of putting unreasonable pressure on themselves to fix things or to enact massive structural changes. Too often, this leads to chronic frustration, as well as feelings of isolation and burnout—when people once passionate about working for a cause grow exhausted, cynical, and detached from it.

In a 2020 study, Panu Pihkala, a researcher at the Helsinki Institute of Sustainability Science, describes the high cost of bearing witness to the environmental crisis. Environmental researchers are at especially heightened risk to suffer from negative psychological impacts of environmental problems because they have knowledge about the scale and severity of environmental problems, they constantly hear about these issues, and they typically have strong emotional ties to the species and ecosystems that are being lost.

This is clearly a problem for the researchers themselves, and, it turns out, for their students. According to Panu, "One of the possible reactions to vicarious traumatization and secondary traumatic stress is that the environmental researchers, both knowingly and unknowingly, pass on the emotions generated by traumatic stress to others."[13] In other words, they inundate their students with doom and gloom, unaware of how destructive this may be. As a university student in an environmental studies program said to me recently, "In the first half of every course you are hit by just how terrible everything is and you look forward to the solutions that will be covered in the second half. But that second half never happens."

To complicate matters, kids who are already overwhelmed by eco-anxiety may experience bullying for caring about recycling or bringing waste-free lunches or protesting pipelines. Youth, as well as teachers, parents, researchers, and activists, say they too are experiencing this kind of environmental backlash. Added to all of this, the urgency, scale, and importance of environmental issues may lead kids and adults to downplay their personal levels of distress. Environmental causes, like so many other critically important issues, can create the discouraging feeling that no matter how much we do, it's never enough.

Be kind

Clearly, in light of these circumstances, engaging in self-care is vitally important. So too is the need for community-building within groups that focuses on appreciating collective strengths and extending compassion to one another. In a 2017 study on kindness, researchers asked 683 adults from two dozen countries, including Brazil, the US, South Africa, and the UK, to engage in four different types of kindness activities. One group practiced self-kindness by going for a walk or taking time to do something else they enjoyed. The second group directed their kindness to close friends and family members. The third group extended kind acts toward acquaintances and people they didn't know very well. The fourth group didn't engage in kind acts, but instead kept an eye out for acts of kindness they spotted others doing. The good news is that kindness in any of these forms increased people's sense of happiness and well-being. As Lee Rowland, lead author of the paper and research affiliate at the University of Oxford, puts it, engaging in and observing kindness is a way of noticing the good around us, rather than seeing a world full of bad news and stress.[14]

Being kind impacts not only our emotions, but our health. Kindness has been proven to help people live longer and better. In 2019, UCLA opened the Bedari Kindness Institute to advance scientific research into kindness and the barriers to it. The Institute's scholars see kindness as an antidote to the polarization, alienation, and isolation plaguing so many places in the world. As Darnell Hunt, dean of the UCLA social sciences division, explains: "I think we're living in a time where there's a direct need to step back and explore the things that make us human and that have the potential to lead to more humane societies."[15]

Cultivate compassion for yourself, other people, and other species

Self-compassion is about being as kind and understanding to yourself when you make a mistake or are having a rough time as you would be to a close friend. This is easier said than done. We are typically much better at giving compassion to others than we are to ourselves. That's because many of us worry that practicing self-compassion is indulgent or a form of self-pity. We fear we will be too easy on ourselves. When we are faced with urgent global issues like climate change, taking time to care for yourself can feel gratuitous. But the reality is just the opposite. Self-criticism turns out to be a poor motivator. It undermines our self-confidence and leaves us anxious and depressed.

Research shows that practicing self-compassion improves our health as well as our capacity to stick with challenging goals, like the big environmental shifts we are trying to make. Self-compassionate people set high standards for themselves. They aren't thrown off when they don't meet a goal and they are more likely to bounce back after a failure rather than sinking into feelings of disappointment. They try hard because they want to learn and grow, not because they think they are inadequate or need to impress others. Self-compassion is not about letting ourselves off the hook. It's about taking care to nurture ourselves so we can achieve our full potential. Kristin Neff, a pioneer in self-compassion research, teaches courses and workshops that recognize self-compassion as a skill we can develop.[16]

Compassion is widely held to be a deep awareness of the suffering of another, coupled with the wish to relieve that suffering. It combines empathy—feeling and understanding someone else's emotions—with action. True compassion means helping people with whom we don't relate. According to James Doty, a neurosurgeon and director of the Center for Compassion and Altruism Research and Education at Stanford University, compassion is what's going to save us from an increasingly fractured world. We need to recognize our common humanity: everyone deserves the right to dignity, food, health care, and shelter.[17]

Lack of compassion for others carries a heavy societal cost. It plays out as stress, anxiety, depression, and bullying. According to the World Health Organization, anxiety and depression cost the global economy $1 trillion per year in lost productivity.[18] Compassion is so important to preventing and tackling these conditions that teaching people how to be more compassionate has been shown to deliver a high return on investment.

Here, again, the necessity of rejecting scare tactics emerges. To be compassionate, people need to feel safe. The biological mechanisms that drive us to nurture and offer care to others are blocked by distress, anxiety, and hostility. "People are wired to pay attention to threat, which triggers fear," James says. "Most things in the world are good; when you create negative narratives, it often leads to other negative events."[19]

In 2019, researchers at the University of Edinburgh announced the discovery of a genetic difference among people who show greater empathy for animals. Previous studies have shown that farmers with higher empathy toward their dairy cattle translate those feelings into the care and attention they provide these animals, which in turn, correlates with higher milk yields. People who share their lives with dogs or cats, and the vets who care for these pets, typically score higher on their capacity to recognize and empathize with an animal in pain. Empathy is thus an important factor in positive relationships between humans and other animals.

It's long been known that our feelings about other animals are shaped by our experiences, cultures, religious beliefs, personalities, and more. This new research represents the first time scientists have shown that genetics may also play a role. Oxytocin is a hormone that boosts social bonding between people. In their experiment, the researchers found that people who showed a higher empathy for animals had a specific version of the gene that produces oxytocin.[20]

Compassionate humpback whales?

Humans may not be the only species capable of compassion. Humpback whales, it turns out, deliberately interfere with attacking killer whales to help others in distress. They don't just defend their own babies or close relatives. They intervene on behalf of other species—a gray whale calf with its mother, a seal hauled out on an ice floe, even an ocean sunfish. Humpbacks act to relieve the suffering of others: the classic definition of compassion.

First-person accounts of animals saving other animals are rare. Robert Pitman, a marine ecologist with the US National Oceanographic and Atmospheric Administration, describes a pivotal encounter he witnessed in Antarctica in 2009. A group of killer whales forced a Weddell seal they were attacking off an ice floe. The seal swam frantically toward a pair of humpbacks that had inserted themselves into the action. One of the huge humpbacks rolled over on its back and the four-hundred-pound seal was swept up onto its chest between the whale's massive flippers. When the killer whales moved in closer, the humpback arched its chest, lifting the seal out of the water. And when the seal started slipping off, the humpback, according to Robert, gave the seal a gentle nudge with its flipper, back to the middle of its chest. Moments later, the seal scrambled off and swam to the safety of a nearby ice floe.

In a 2016 article in *Marine Mammal Science*, Robert and his coauthors describe this behavior and confirm that such acts of do-gooding are surprisingly widespread. They have been occurring for a long time and have been seen in locations all over the world. "Now that people know what to look for, especially people out on whale watch boats, they see it fairly regularly," Robert says. "So now, even for the people who didn't believe, which initially included some of the coauthors on the paper, I think everybody now understands that this is going on."

Intriguingly, humpbacks don't just stumble upon killer whale attacks. They race toward them like firefighters into burning buildings. And like these brave rescue workers, humpbacks don't know who is in danger until they get there. That's because the sound that alerts them to an attack isn't the plaintive voice of the victim. It's the excited calls of the perpetrators. Transient killer whales tend to be silent when they are hunting, but when they finally attack they get really noisy. Robert believes humpbacks have one simple instruction: "When you hear killer whales attacking, go break it up."

Steve Cole, a professor of medicine and psychiatry and biobehavioral sciences in the UCLA School of Medicine, reveals an intriguing insight into threat biology that might shed further light on why humpbacks actively enter into dangerous altercations with killer whales. Steve explains that scientists used to think that the circuitry for detecting and responding physiologically to threatening circumstances was there to protect the survival of the individual. But that is no longer the case. Studies in threat neurobiology suggest that those circuits are there to defend the things individuals care about. "This is why you get parents and those firefighters running into burning buildings to save children, and soldiers running into a hail of gunfire for the country they love," says Steve.

So, are humpbacks compassionate? When I pose this question to Robert he says, "No science editor is going to let me use the word compassion. When a human protects an imperiled individual of another species, we call it compassion. If a humpback whale does so, we call it instinct. But sometimes the distinction isn't all that clear."

We now recognize cultural differences within whale, primate, elephant, and other species in ways that were unimaginable just decades ago. Studies of animal emotions proliferate, and with them come challenging questions about how best to interpret what looks like compassion and altruism in other species. Just how these acts differ from our own behaviors may be hard to pinpoint. In 2014, for example, commuters in a crowded railway station in northern India watched a male rhesus macaque attempt to resuscitate an unconscious macaque that had been electrocuted while walking on high-tension wires. A video of the incident shows the rescuer nipping, massaging, shaking, and repeatedly plunging the victim into water. The life-saving effort lasts twenty minutes, until the monkey miraculously revives.

In an attempt to decipher what qualities of compassion might be uniquely human, I binge-watch videos on Stanford

University's Center for Compassion and Altruism Research and Education website. I am captivated by a video showing a series of experiments in which a toddler voluntarily totters across a room to assist an apparently clumsy researcher who needs help. The same basic helpful behavior happens later in the video when the experiment is repeated with chimpanzees.

What's powerful about these studies, according to Felix Warneken, principal investigator of the Social Minds Lab at the University of Michigan, and the researcher who led the study, is that they challenge the strongly held belief that we need to be taught to be altruistic through social norms. His findings indicate otherwise. Chimpanzees, as well as children too young to have learned the rules of politeness, spontaneously engage in helpful behaviors, even when they have to stop playing or overcome obstacles to do so. The same results have been duplicated with children in Canada, India, and Peru, as well as with chimpanzees at the Max Planck Institute for Evolutionary Anthropology in Germany and other research centers across the world. The chimps helped not only people they knew, but human strangers too.

COMPASSION, IT TURNS out, is innate. Human beings and other animals have what Dacher Keltner, a professor of psychology at the University of California, Berkeley, calls a "compassionate instinct."

All compassion involves some benefit for the helper, Steve Cole says. In fact, the happiness we derive when we act on behalf of the greater good shows up in our cells as a bolstered immune response, he says. While we might feel just as happy eating ice cream as we do volunteering at a beach cleanup, at a cellular level happiness that comes from meaningful service to others is correlated with positive health benefits. Better health through taking action on behalf of the collective—that's good news indeed for engaging people in fixing the Earth.

8

TRENDING HOPEFUL

The asteroid that hit the Earth 66 million years ago
triggered a massive 11-magnitude earthquake
that, in turn, set off a wave of volcanic eruptions
that lasted 300,000 years.

These seismic shifts
killed off the dinosaurs
and gave birth
to hydrothermal vents
filled with fish and snails and tubeworms
who thrive in oven-hot temperatures
completely in the dark,
giving birth in sulfuric acid
at the bottom of the ocean.

Everywhere I look
I see the impossible
made possible.

A DECADE AGO, I felt utter despair over the catastrophic rise in the murder rate of elephants. Between 2007 and 2014, 30 percent of savanna elephants (144,000 animals) in Africa died, primarily at the hands of poachers.

This horrific situation is made even sadder by the fact that elephants are such remarkable animals. They're highly intelligent, social, caring, and they're deeply attached to members of their extended families. In response to losing family members in such a violent way, elephants suffer from the psychological symptoms of post-traumatic stress, depression, unpredictable aggression, and antisocial behavior, according to a heartbreaking 2014 study published in *Nature*.[1]

People have been decrying elephant poaching for decades, but the complexity of the issues that drive it—poverty and lack of sustainable livelihoods for people living in close association with elephants in African· countries; widespread corruption; and a deep history of ivory carving as a cultural craft and status symbol across Southeast Asia, especially China—stoked an insatiable demand for elephant tusks.

In 2014, Microsoft co-founder Paul Allen launched the Great Elephant Census project. Ninety researchers covered

285,000 miles (459,000 km) across twenty-one African countries, creating a pan-African aerial survey and a massive raw data set. The resulting census map provides a detailed analysis of emergency situations where elephants face local extinction or staggering declines, and other areas where populations are increasing.

It would be naive to suggest that big data allows us to know better, and that "knowing better" on its own would stop poaching. That isn't the case. But having more robust evidence, and transparency to see patterns, trends, and hotspots, triggered meaningful change. The Great Elephant Census project didn't start or end with this much-needed map. It's also established relationships within collaborative networks to fight against poaching in Africa and the illegal trade chains in Asia.[2]

Recognizing that saving elephants can't happen without the support of Chinese consumers and the Chinese authorities, a massive public-awareness campaign was launched, including a major campaign by Yao Ming, the superstar basketball player. The Guardian newspaper teamed up with the media outlet chinadialogue to provide a year of in-depth reporting on the elephant crisis in Chinese languages.[3]

Attitudes within China began to shift. When surveyed, 95 percent of respondents in Beijing, Shanghai, and Guangzhou (China's three biggest cities) revealed that they wanted the government to ban the ivory trade in order to protect African elephants.[4] Meanwhile, international pressure on China from African and European countries and the US, along with the World Wildlife Fund, INTERPOL, the UN, and many other organizations, intensified.

On December 31, 2017, China officially made all trade in ivory illegal. It was a huge step in the right direction, and one that is yielding results. Elephant poaching in Africa has dropped dramatically, according to a 2019 report in *Science*.[5]

While big data enabled a detailed analysis of the problem, collaborative networks brought together vital expertise to understand the root causes and create relationships for coordinated responses. In this era of knowledge-driven economies, big data and collaborative networks are dramatically changing the way we understand and can respond to environmental crises.[6] Indeed, we now know so much about the global threats of climate change in large part thanks to the embrace of big data in scientific research.[7]

Trends determine the food we eat, the clothing we wear, the way we work, the music we listen to, and so much more. Yet we rarely think about trends when it comes to the environment. We need to change that. We need to bravely look at the troubles we face and then situate our environmental actions as part of timely movements. When something is trending, it generates tremendous momentum. Seizing upon that momentum and amplifying it changes things fast.

Environmental issues are hugely complex and daunting. The world is filled with devastating problems and worrisome trends. It's almost impossible not to get overwhelmed by the magnitude, details, and sheer number of problems that need addressing. That's why paying attention to trends that are heading in positive directions for the planet is hopeful. These trends show us where and how collectives of people with expertise, energy, influence, and guts are achieving meaningful results. Like strong currents, they carry us forward, helping us ride out the inevitable setbacks and soul-crushing obstacles that accompany transformations. Remember, emotions are contagious. The more you engage with trends that are achieving meaningful results, the more hopeful you feel, and the more you spread those feelings to others who will, in turn, amplify transformative solutions.

Trend: Hello climate emergency declarations

An astonishing surge of "climate emergency" declarations happened in synergy with the 2019 climate marches. More than 1,261 jurisdictions in twenty-five countries signed on.[8] Cities, associations of scientists, religious groups, major companies, whole countries, the EU—on the eve of 2020, one in ten people on the planet lived in a place that had declared a climate emergency.[9]

While it might seem counterintuitive to see these declarations as a positive trend, they represent a staggering show of global support for tackling the climate crisis. Cities, especially, are using that momentum to drive change. That's important because more than half of the world's population lives in urban and suburban areas. Cities are where emissions are the largest: they account for about 70 percent of carbon dioxide emissions, according to a 2018 report released at the IPCC Cities and Climate Change Science Conference.

In October 2019, mayors of the C40 Cities, a network of ninety-four of the world's largest cities, announced that thirty of them—London, Los Angeles, Montreal, and Warsaw, to name a few—have already brought their greenhouse gas emissions in line with science-based targets to limit global temperature rise to 1.5 degrees Celsius (2.7°F). In other words, 58 million urban citizens now live in cities that are steadily reducing their greenhouse gas emissions. These cities have already met a critical milestone of the Paris Accord.[10]

Leading the pack is Copenhagen. This capital city of 600,000 people is on track to becoming the world's first 100-percent-carbon-neutral capital by 2025—while cutting prices for energy users. They've already dropped carbon emissions by more than 40 percent since 2005. They've switched to wind energy and other renewable sources, and will replace their coal-fired power plant with biomass-powered units. By July 2020, the most polluting gas and diesel trucks and vans will not be allowed to drive in the city.

The speed at which Copenhagen is moving to decarbonize provides encouragement for other cities. It provides real-world corroboration for what Paul Romer, an economist at New York University, said at the press conference announcing his 2018 Nobel Prize for his work on the economics of innovation and climate change: "It's entirely possible for humans to reduce carbon emissions," he said. "There will be some tradeoffs, but once we begin to produce [fewer] carbon emissions we'll be surprised that it wasn't as hard as it was anticipated."[11]

Even in countries where climate *inaction* at the national level is rife, powerful actors are pressing forward to reduce carbon emissions. We Are Still In is a network of 3,850 American businesses, states, cities, faith groups, universities, museums, environmental organizations, and other entities that have joined together to continue their efforts to achieve the greenhouse gas emissions goals of the Paris Agreement. Together, they are responsible for nearly 70 percent of the country's gross domestic product and half of its population. If We Are Still In were a country, it would be the third-largest economy in the world.[12]

Similar initiatives are emerging in countries such as Vietnam, Argentina, South Africa, and Mexico, where local and regional governments are joining with businesses and climate activists to cut greenhouse gases, often in clear opposition to their national governments' plans to increase investments in fossil fuels.

The role of businesses in these networks marks a noteworthy shift from a decade ago, when climate action and economic growth were often viewed as being in conflict. A 2018 report by the Global Commission on the Economy and Climate, an international commission made up of former heads of government, finance ministers, and leaders in business and economics, found that bold climate action could generate more than $26 trillion in benefits through 2030.[13]

Trend: Bye-bye, single-use plastic

A rush of positive activity is happening around single-use plastic too. I sometimes serve as a volunteer judge at student science fairs, and I often encounter projects on plastic pollution done by children who have seen images of sea turtles choking on plastic bags. Many of them are among the 35 million viewers of a particularly gruesome YouTube video of a sea turtle with a plastic straw being removed from its nose. These kids feel overwhelmed by the horror of plastic pollution and their inability to stop other people from using bags, straws, and other single-use plastics.

Their concern is well-placed. Plastic makes up 80 percent of all ocean debris, and plastic bags are often described as the most ubiquitous consumer product on the planet.[14] Yet think how empowered these kids would feel if they knew their science fair projects were part of a global movement that is causing the downfall of the plastic bag.

More than 127 countries have already imposed restrictions and bans on plastic bags, and even more action is happening at the level of cities, provinces, and states. The African continent leads the world in bag regulations, with thirty-four countries adopting taxes or bans. Thirty-one of these countries are in sub-Saharan Africa, the world's poorest region. Kenya's penalties are the world's most punitive, with manufacturers, importers, distributors, and users facing up to $38,000 in fines or four years in prison.[15]

Researchers say countries in the Global South, where clumps of plastic bags create breeding grounds for malaria-bearing mosquitoes and clog drainage systems, are leading the way with bans and stiffer penalties, whereas those in the Global North lean more heavily toward levies, taxes, and fees.[16] In Denmark, which passed the world's first bag tax in 1993, residents use, on average, four plastic bags per year.[17]

Exceptions for certain kinds of plastic bags exist, even in places with bans. This may cause you to wonder if plastic-bag initiatives are more greenwashing than substance. But measures to target single-use plastic bags are yielding impressive results. For example, CalRecycle, California's state recycler, reported an 85 percent reduction in the number of bags used after California passed a statewide ban on plastic bags in 2016. Similar-sized drops occurred in Wales, Northern Ireland, England, and Scotland when major stores in the UK began charging the equivalent of six cents per bag. In a 2019 study, researchers revealed that people in all age, gender, and income groups in England substantially reduced their plastic-bag usage within a month after the charge was introduced. Even better, support for the plastic-bag charge created a spillover effect, increasing public support to reduce other forms of plastic waste.[18]

Bans on straws, cutlery, plates, and more are spreading quickly. In October 2019, the European Parliament overwhelmingly voted in a ban on single-use straws, stir sticks, cotton buds, and cutlery by 2021. They set higher recycling targets and strengthened the polluter pays principle—which means, for example, that manufacturers and not fishers will incur the cost of collecting nets lost at sea.[19] Fishing nets are a major source of ocean plastic pollution, accounting for almost half of the plastic in the Great Pacific Garbage Patch.[20]

Meanwhile, a growing range of brands are using recycled ocean plastic in their products. Bureo makes sunglasses and skateboards from recycled fishing nets. Net-Works, a collaboration between the Zoological Society of London and Interface Inc., a global carpet tile manufacturer, works with communities in the Philippines and Cameroon (and soon Indonesia) to collect and sell discarded fishing nets. Yarn made from the recycled nets becomes carpet tiles, and the entire system supports sustainable livelihoods. Tens of millions of pairs of Adidas shoes made of more than 40 percent recycled ocean plastic have sold since 2015.

When China stopped importing plastic waste in 2018, it inadvertently did the world a favor. Prior to that date, two-thirds of the world's plastic waste was shipped there. Beijing's decision to halt shipments of certain plastics, paper, and textiles meant that plastic waste has shifted from being a global problem to a country-of-origin problem. It has already sparked new investments in recycling in the US and other countries.[21]

Trend: Goodbye food waste

If you haven't seen Project Drawdown, it's time to take a look. This global research organization identifies, reviews, and analyzes the most effective solutions to climate change. It ranks "reducing food waste" as the third most important step to tackle climate change, out of a list of eighty solutions. Recovering just 25 percent of wasted food could effectively end world hunger.[22]

The good news is momentum against food waste is spreading fast. In addition to the apps and other personal technologies we discussed in chapter five, international efforts are under way. Australia and Norway have committed to halving the food waste they produce by 2030. More than 90 percent of farming in Indonesia, the fourth most populated country on the planet, is done by small family farmers. They stand to benefit directly from an international program launched in 2019 to reduce food waste by 50 percent in the next decade.[23] At the international level, the United Nations Environment Programme and the Food and Agriculture Organization support a global alliance that is actively working on this issue. The European Union's Environment Committee has pledged to halve food waste by 2030.

In 2016, France made it illegal for large supermarkets to throw away unsold food. Stores must instead donate the food to charities. According to the French Federation of Food Banks, more than half the food given out by the five thousand charities in its network now comes from grocery stores, and the quantity and the quality of food has improved. In 2019, France amplified this successful approach, extending the ban to include non-food items like clothes, shoes, cosmetics, textiles, electronics, and other products. According to ecological transition minister Brune Poirson, the scheme will ensure that a billion euros worth of unsold non-food items will be donated or reused rather than thrown away or destroyed.[24]

The Finnish supermarket chain S-Market hosts a "happy hour"—reducing food prices by up to 60 percent an hour before closing for items close to their expiry date. After the store closes, the remaining discounted products are donated to more than eighty charities, and any inedible leftovers are converted into biofuel. Meanwhile, Italy provides tax breaks to supermarkets that donate unsold food. The Danish supermarket Wefood goes even further: its entire stock comprises waste-food items, which it sells at greatly reduced prices. It has expanded to three stores since 2016.

Food-industry leaders are also tackling the waste issue. In 2019, eight leading Canadian companies, including Kraft Heinz Canada, Walmart Canada, Loblaw Company, Save-On-Foods, and Sobeys Inc., formally committed to drop food waste in their own operations by 50 percent by 2025, and to measure and report on their progress.

Banning food waste from landfills is a growing trend across states, provinces, and municipalities in North America, and around the world. With cities now home to more than half the world's population, and typically having authority over their own municipal waste programs, they have become powerful leaders in food rescue programs, curbside composting, and other innovations in the waste-food-reduction movement. In the lead-up to the 2015 World Expo in Milan, Italy, the mayor of Milan initiated the Milan Urban Food Policy Pact. This growing network of two-hundred-plus cities from around the world represents 450 million inhabitants, and they've signed on to create sustainable food systems and reduce food waste.

South Korea, for instance, now recycles 95 percent of its food waste. In the capital city, Seoul, residents deposit food waste in biodegradable bags into pay-as-you-recycle smart bins equipped with scales and radio-frequency identification that charge based on weight. The bags are compressed in a processing plant: the moisture is used to create biogas, and the dry waste is converted into animal feed and fertilizer used in Seoul's rapidly growing urban-farm movement. The number of community gardens popping up on the roofs of schools and office buildings and between apartment blocks in that city jumped sixfold between 2012 and 2019.[25] Research on the impact of community gardens around the world reveals that growing food together not only improves our diet and health but also lowers stress, increases our physical activity, enhances our sense of belonging, and strengthens our relationships to other people in our communities.

Trend: Hello livable cities

Fossil fuel combustion is the world's most significant threat to children's health, and it's a major cause of global inequality and environmental injustice.[26] In 2018, the Lancet Commission on pollution and health found pollution to be the largest environmental cause of disease and premature death. Each year, pollution kills fifteen times more people than all wars and other forms of violence combined.[27] Both air pollution and climate change demand a rapid transition to more sustainable energy models. Changing the way we power our cities and how we travel around and between them is crucial. What's exciting is that the innovations being put in place to solve these issues have benefits not only for addressing climate change and pollution, but for increasing our joy and sense of community too.

A global mass transit revolution

When we walk, cycle, or use public transit, we improve our own health, lower health care costs,[28] lower greenhouse gas emissions, reduce air pollution, and reduce other environmental and safety problems caused by cars. We're undergoing a global mass-transit revolution. The world is building mass public transit networks faster than ever before, and the number of people switching to use them is increasing. In 2017, 53 billion passengers traveled by mass transit. That's an increase of about nine billion passengers since 2012.[29]

Bikes, bikes, bikes

Cities are turning to bikes to help solve congestion, pollution, climate change, and safety and health issues. Today nearly a thousand bike-sharing schemes exist in cities. Fifteen years ago, there were four. Indeed, between 2014 and 2018, the number of bike-sharing programs around the world doubled, and the number of public bikes increased by a factor of twenty.[30] And when we make cities better for bikes by calming traffic, adding bike lanes, and ensuring cycling access routes cover the city and join up with public transit centers, it also has unexpected benefits. Comprehensive bike- and road-safety studies demonstrate that building safer routes for cyclists is one of the biggest factors in making streets safer for everyone, including car drivers. By calming traffic and slowing speeds, bike infrastructure reduces the rate of fatal car crashes, which kill forty thousand people every year in the US, according to a 2019 study.[31] With added bike lanes, fatal car crash rates dropped in Portland by 75 percent, Seattle by 60 percent, Denver by 40 percent, and Chicago by 38 percent. Fortaleza, the fifth-largest city in Brazil, saw traffic fatalities drop to their lowest level in fifteen years after creating cycling infrastructure, priority bus lanes, and lower speed limits on city streets.

Peak car?

With more modes of mobility to choose from—think scooters, bike sharing, public transit, ride-hailing, and car-sharing—increasing concerns about climate change, air pollution, and gridlock, and the proliferation of mobile apps that can beckon a vehicle on demand, private car ownership for many people is becoming obsolete, especially among people who live in cities where cars can be an expensive inconvenience. This is a crucial trend, since two-thirds of the world's population is expected to live in megacities by midcentury. (Almost 90 percent of urban population growth in the next four decades will be in Asia and Africa.)[32]

Millennials and Gen Z in particular are turning away from car ownership. Surveys reveal many reasons they eschew cars—cars are a major cause of air pollution and greenhouse gas emissions, the horrific number of people killed by cars, the expense and inequity of car ownership, and the headache of maintenance and parking, to name a few. About 64 percent of India's Gen Z and millennials question the need to own a vehicle, and 59 percent of them are now using ride-hailing services, up from 38 percent just two years before, according to the 2020 Global Automotive Consumer Study.[33] The decline in young people getting driver's licenses is now a trend in Canada, the US, Sweden, Norway, Australia, the UK, Germany, and other countries.[34]

A number of big cities are getting serious about congestion pricing, one of the biggest positive steps to reducing the number of cars in city centers. It works by charging cars to drive in popular areas. The result is fewer cars on the road, which means less pollution and traffic jams. Singapore has been using it for almost fifty years. When London, England, began congestion pricing in 2012, traffic dropped by 30 percent in the first year. New York City will be the first US city to impose congestion pricing starting in 2021, when drivers will have to pay to enter the core of Manhattan. The money raised will be used to modernize the public transit system. Other North American cities are projected to follow suit.

Taking back the streets for pedestrians

A growing number of cities are turning streets back over to people. In the city center of Paris, the first Sunday of the month is car-free day. A French air-quality-monitoring network reported a 40 percent decrease in exhaust emissions on the day of the inaugural event.[35] Meanwhile, Oslo, Norway, removed the last car-parking spaces from its downtown core in 2019, creating a car-free city center. Pedestrian use jumped 10 percent in a year.[36]

The year 2010 marked the beginning of a transition from car-centric to people-friendly for Bucheon, South Korea. Treed and flowered walking trails now connect the city's 188 libraries. Small libraries were created in subway stations, and a system was developed to deliver books to people with mobility challenges. The number of pedestrians using the green spaces and walking trails increased to four million, and Bucheon is now recognized as a UNESCO Creative City of Literature.[37]

Barcelona, Spain, is building car-free superblocks, while Germany has started work on an "autobahn for bikes" that will create a network of car-free corridors across ten cities and four universities. In 2019, when the new mayor of Madrid, Spain, repealed a car ban that was already in place in the city center, air pollution rose sharply and thousands of people took to the streets to protest. Support for cleaner air and putting pedestrians first was so strong, the low-emission zone was reinstated five days later.[38]

Trend: Welcome urban
forests and wilder neighborhoods

Increasingly, city planners and officials are turning to a new ally to tackle big city problems—trees. Urban forests make cities more beautiful, plus they reduce air pollution and protect us from harmful effects of the sun. Throughout their lives, trees take up carbon through photosynthesis, which fights climate change and gives us healthy air to breathe. Urban forests in Ouarzazate, Morocco, combat desertification as well as clean wastewater. Fruit from city trees supplies food banks in Seattle. Greenbelts of trees in Bobo-Dioulasso, Burkina Faso, help to cool city areas experiencing higher temperatures from the energy created by so many people, cars, buses, and buildings.

People involved with public health often look at the social determinants of health—specific conditions such as access to housing, fresh food, public transportation, education and work opportunities, and medical clinics, as well as air quality, crime rates, and other neighborhood characteristics that affect our health and quality of life. Urban areas tend to concentrate poverty and poor health. A 2018 study conducted by the Social Science Research Council reveals a shocking thirteen-year difference in life expectancy depending on where people live in New York City; residents of Bayside, Douglaston, and Little Neck in Queens have a life expectancy of 89.6 years, compared to a 76.7-year life expectancy for people living in Brownsville and Ocean Hill in Brooklyn.[39]

Thanks to projects like MIT's Treepedia, which uses data from Google Street View to map the trees in cities, it's easier than ever to compare the amount of tree cover in different neighborhoods, which can help illuminate disparities and spur political action. In more than 130 countries, people use i-Tree, a software suite from the USDA Forest Service that provides assessment tools and analysis to help them select the best types of trees for their area and the optimum places to plant them.

Urban forests improve our lives and make good money sense, too. According to a 2018 study, every dollar invested in urban trees provides a return on investment of $2.25 in terms of cleaner water, climate action, health and well-being, and more.[40] In another study, researchers determined the economic benefit of London's urban forest and then extrapolated it across ten other megacities on five continents, including Mumbai, Buenos Aires, and Los Angeles. Each year, the tree-based ecosystem benefits per megacity had a medium value of $505 million.[41] By planting more trees in degraded areas, they could increase these benefits by 85 percent.

Trees positively impact how we feel and support better health outcomes. A study in Toronto found having ten or more trees in a city block improved how people rated their health—by a level comparable to an increase in annual income of $10,000 or being 1.4 years younger.[42] Trees also help make cities safer. Studies in Baltimore, Portland, and Philadelphia demonstrate a strong association between treelined streets and a substantial reduction in crime. When Philadelphia cleaned up vacant lots and planted trees in them, gun assaults across most of the city dropped by 8 percent. In Baltimore, a 10 percent increase in large tree cover correlated with a 12 percent decrease in crime.[43]

Urban forests are just one example of a growing momentum to address city problems like air pollution, flooding, carbon emissions, and more by restoring and conserving biodiversity.[44] More and more studies demonstrate how investing in restoring, protecting, and enhancing green infrastructure in cities is economically, socially, and ecologically wise.[45] Time spent in nature is now proven to carry important physical, mental, and social health benefits. Epidemiological evidence details how living in greener city areas is associated with lower probabilities of diabetes, cardiovascular disease, obesity, asthma, stroke, and mental health issues, and with better cognitive development.[46]

It's a hopeful feedback loop: according to researchers, people who spend more time in nature, including hanging out in city gardens or engaging in urban birding, grow attached to the places they frequent and become more likely to push for projects to conserve them.

Trend: Hello solutions

The science of bright spots

A growing number of researchers now investigate "bright spots": places where a species or ecosystem is doing far better than one might expect given the stresses they face. The goal is to understand what they're doing right. In a massive global study of six thousand coral reefs across forty-six countries, a team of thirty-nine scientists from thirty-four universities revealed that conservation is possible, even in areas heavily used by people.

Josh Cinner, the lead researcher on the reef study, got the idea of looking for positive outliers from the field of public health. Thirty years ago, Jerry and Monique Sternin went to Vietnam to fight malnutrition on behalf of the charity Save the Children. They set out to find children who were bigger and healthier than average, despite suffering the same levels of poverty and disadvantage as their neighbors, and to figure out why. They discovered that the mothers of the better-nourished kids did two things differently. The mothers divided the food into smaller portions so they could feed their kids four times a day rather than the norm of twice a day, and they supplemented that food with things they could gather or forage in their area, like tiny fish, shrimp, and sweet potato greens. By identifying what was working and then teaching it to other mothers in the village, the Sternins helped cut child malnutrition by 65 percent. Eventually 2.2 million people across 265 villages benefited from the program.[47]

By adopting this proven approach of looking for positive outliers or bright spots, Josh and his colleagues discovered that strong local involvement in how reefs were managed and high dependence on these fisheries were important. Community ownership rights to reefs enabled people to commit to and develop creative solutions to problems.[48]

Solutions-oriented environmental assessments

Scientific studies from a large and diverse range of fields are analyzed by large groups of scientists and other experts to create global environmental assessments (GEAs) of complex issues, which are used to inform policy makers. The updated assessments issued by the Intergovernmental Panel on Climate Change are a familiar example of a GEA. These assessments built the evidence base that tells us we are in a climate crisis.

The climate crisis is here. As policy makers, political leaders, and decision-makers of all kinds increasingly seek evidence-based advice on the best ways to tackle it, GEAS are transforming to meet that demand. Increasingly they are providing not only problem analyses but also rigorous analyses of specific solutions.[49] It's now also easier to source this advice, thanks to global research programs like Future Earth, which draws on expertise, networks, and research from around the world to generate solutions for complex issues.

The rise of solutions journalism

In the burgeoning field of solutions journalism, reporters bring the same rigorous reporting skills they apply to covering societal problems to the investigation of what's working: what interventions have the greatest success rate in tackling the opioid epidemic, or proven approaches to prevent school shootings, for instance. Solutions journalism reveals ways people are responding to crises and focuses on effectiveness and outcomes, not just good intentions. Rather than advocating for a specific solution or approach, solutions journalism reports on responses others are using to tackle the problem that is being investigated.

Solutions journalism is quickly spreading through newsrooms and journalism schools worldwide in print, digital, and now television media, thanks to the creation of the Solutions Journalism Network. Since its inception in 2013, its "Solutions Story Tracker" has amassed more than eight thousand stories produced by over a thousand news outlets across 165 countries. It's a brilliant resource for searching for solutions-oriented stories on any subject you can imagine.

The turn toward reporting on solutions is further fueled by the rise of digitally born media. The majority of Gen Z get their news from digital-first publications such as BuzzFeed, Vice Media, Business Insider, Huffington Post, and Quartz.[50] These media platforms, in general, have invested more heavily in covering the environment and climate change. They are more likely to focus on solutions than legacy or traditional media, according to the Reuters Institute for the Study of Journalism.[51]

Solutions journalism is forward-looking. It moves beyond the familiar role of exposing wrongdoing: detailing the problem, of course, but then rounding out the story by including the innovative ideas and pathways people are already using to solve or fix what's broken, and reporting the evidence of what results are being produced. In so doing, solutions journalism stories don't just reassure people; they let them know what they can do—and how. People are doing remarkable things all over the world, but with the media overwhelmingly focused on corruption, scandal, and disaster, these positive events too often go unreported.

Problems often persist because people don't know what they can do to address them more successfully. By bringing real-world solutions into view, solutions journalism plays a key role in counteracting the destructive cynicism and distrust I described in chapter two. Solutions journalism also helps hold those in power accountable to make change. For instance, a 2019 story in *Popular Science* about how Flagstaff and Tucson, Arizona, have implemented city-wide changes to dramatically cut light pollution, which harms people and other animals, provides evidence others can use to demand similar actions in their own cities.

Covering Climate Now is an initiative to create more and better coverage of climate change issues and solutions, at local to international scales and across a broad range of media. Launched in April 2019, it has rapidly grown to include hundreds of media outlets from around the world, reaching a combined audience of more than one billion people. It's the largest media project ever organized around a single topic.

The birth of #OceanOptimism

In the mid-2000s, when my nephew, Matthias, was around ten years old, he asked me, "If everyone is doing all this stuff for conservation, has anything gotten better?"

It was such a good question, and yet it quickly became apparent, as we dug through the library and looked online, that the information that would help us answer it was surprisingly difficult to find. His question set me on a solutions-finding spree, interviewing scientists wherever I was in the world and asking them to share their best examples of hopeful environmental successes that were backed by scientific evidence. Two things struck me with their answers. The first was that they had no shortage of examples of positive change, and the second was that many told me I was the first person to ask them that question. It made me realize that looking at the environment as an endless series of problems to be solved was a taken-for-granted norm. So much so that asking about solutions surprised them.

The more solutions I collected, the more I wanted to share them, but I soon realized there was no obvious place to do so. I started writing about them in magazines, adding them to talks I gave, and I wrote *Not Your Typical Book About the Environment*, which is a hopeful book for kids based on scientific evidence. I thought at least I would have a book that I could share with kids who were asking that same good question as Matthias. Yet I knew these small efforts weren't nearly enough. The more I researched the impact of environmental doom and gloom, the more I felt compelled to increase access to real-world examples of conservation successes to counterbalance that narrative.

In 2012, I reached out to Nancy Knowlton of the Smithsonian Institution, Heather Koldewey of the Zoological Society of London, and Cynthia Vernon of the Monterey Bay Aquarium, three powerhouses in ocean conservation who I had discovered shared a passion for increasing access to ocean successes. I simply invited them to my house in Pacific Grove, California, for the weekend, and they generously agreed to come.

Earlier that year, I had been lucky enough to be a writer-in-residence at Hedgebrook on Whidbey Island, where I was introduced to the concept of "radical hospitality." The idea is to nurture people's most creative selves by taking care of all of their needs for delicious food and warm beds and to give them freedom to use time as they might wish. I happened to be doing a book talk at my local library a few days before Nancy, Heather, and Cynthia were due to arrive, and

I must have briefly mentioned wanting to host them in the spirit of radical hospitality—and, as I was then a terrible cook, how nervous I was about my capacity to create the "delicious food" part of the equation. I thought nothing more of it. But that weekend, just as we were settling into our discussions, there was a knock at my front door. A half-dozen folks who'd been at the library talk walked in carrying homemade soup, freshly baked bread, and warm-from-the-oven cookies.

This unexpected act of generosity brought the spirit of radical hospitality to life. We felt supported and inspired to share ideas and to really get to know one another as we walked the beaches of Monterey and chatted on my front porch. I am convinced that generous beginning helped us to move our agenda forward even though we lived and worked in different countries.

Heather's desire to source and share hopeful marine solutions arose from her concern about the tendency for scientists to publish problem analyses rather than conservation successes. Heather travels extensively in her role as head of the Zoological Society of London's marine and freshwater conservation programs. She frequently encounters marine conservation practitioners working in isolation without access to proven approaches.

Nancy's interest in focusing on hopeful solutions stemmed from witnessing the impact of doom and gloom on the graduate students she taught, and on the field of marine science more broadly. "An entire generation of scientists has now been trained to describe, in ever greater and more dismal detail, the death of the ocean," she wrote in an article with her husband, the noted marine scientist Jeremy Jackson. In an attempt to balance that view, Nancy hosted what she called "Beyond the Obituaries" sessions at major international science conferences. Scientists were invited to only share conservation success stories. She thought a few people might show up. To her surprise, the sessions were packed.

In 2014, with the help of Elisabeth Whitebread, a global marine community organizer, we gathered a small group of creative people together in a little village on the outskirts of London, England. We challenged ourselves to use the forty-eight-hour workshop to create and pilot a social-change project to engage people with ocean conservation successes and shift the environmental narrative beyond doom and gloom. The result was #OceanOptimism—a social media campaign that crowdsources and shares ocean solutions and successes that are currently happening all over the world. We launched it on World Oceans Day, June 8, 2014, and it has reached more than 90 million shares to date.

Following the success of #OceanOptimism, Nancy Knowlton has led the Smithsonian Institution in a massive Earth Optimism initiative that includes an international summit with sister events spread around the globe, as well as an ongoing #EarthOptimism social media campaign. Other organizations and groups are rapidly following suit. The University of Oxford now hosts #ConservationOptimism. And influential thought leaders, including Jane Lubchenco, University Distinguished Professor at Oregon State University, are calling for new solutions-based narratives for the environment.

A significant barrier to feeling hopeful about the future and amplifying solutions is lack of easy access to what's working. Today, with a simple click on #OceanOptimism, I am awash in uplifting news of marine conservation successes: a new marine sanctuary in the Galápagos Islands to protect the world's highest concentrations of sharks; an "unprecedented" comeback for Antarctic blue whales; green sea turtles in Florida and Mexico no longer listed as endangered, thanks to successful conservation efforts; a no-take zone that revived a Scottish fishery devastated by dredgers.

We are *not* at the starting line

I wish I could share a complete list of hopeful environmental trends, but that list is too vast and ever-growing to capture. Too often our language around climate change and other environmental issues is "if we do this, then we have a chance of this." The future orientation of the phrasing feeds a mistaken impression that nothing has been done. It creates the daunting sense that all of the hard work lies ahead. I consciously try to counteract this starting-line fallacy by talking about the environment in the present tense, and positioning current issues within a trajectory of past accomplishments— *Monterey Bay is healthier now than it has been in the past two hundred years thanks to all the people who pushed for marine protected areas, fishing regulations, the Marine Mammal Protection Act, and other conservation measures across many decades, and that's why it's so important to support the rapidly growing global trend to ban single-use plastic.*

Solutions are not final, perfect end points. They're ongoing processes that require monitoring and adjustment to achieve meaningful results. Solutions are directions that require constant vigilance. But the need for vigilance shouldn't prevent us from forward action. A solutions orientation to the climate

crisis requires us to welcome the inevitability of making mistakes because we know that going down the wrong path on
occasion is an essential part of any new creative, collaborative venture.

By moving in the direction of what works and creating a trusted, evidence-based feedback loop of the impact
of positive changes in fishing practices, the Monterey Bay
Aquarium's Seafood Watch program is revolutionizing the
sustainable seafood industry. In the twenty years since it
started it has grown to include 180 conservation partners
across seven countries. More than 100,000 business locations
in North America rely on Seafood Watch science to inform
their purchasing decisions. Eighty percent of the seafood
consumed in the US is eaten in restaurants.[52] The Seafood
Watch app provides people with a list of sustainable seafood
to choose and others to avoid. Giving us the power to make
informed choices about how and where specific fish and other
seafood is caught rewards good practices. It's such an effective agent of change, the app is updated once a month. The
global sustainable seafood market has grown to more than
$12.71 billion and is expected to reach $18.63 billion by 2025.[53]

WHEN WE SITUATE ourselves in well-informed networks
and amplify what's working, we overcome formidable odds.
Hope exists within the agency of the countless other animals,
plants, and other life-forms that populate this remarkable
planet. It exists within our proven collective capacity to enact
meaningful change. Hope is wild and contagious. My wish
is that you will nurture the wild contagious hope that lives
inside of you and actively spread it to everyone you know.

AFTERWORD

T A TEACHERS' workshop on hope and climate change, one of the participants shared his grief about what is happening to his seven-year-old daughter. She's so overwhelmed, she's taken to gathering plastic waste and stockpiling it in her room; falling asleep in tears with single-use bags stuffed beneath her bed. *What can I do?* her father asked us. It was such a poignant example of the damage of doom and gloom; we responded with silence.

Listen to how she is feeling, one of the workshop participants quietly offered. *Share how you are feeling with her. Show her a map of all of the places that are banning plastic bags so she can see how many other people are helping to solve this problem too,* someone else said.

How else might we compassionately support this little child? Here is a list I started. I hope you will add to it, and modify it to nurture a special person in your life.

Remind her that situations can get better by pointing out how the bald eagles that are now plentiful in her neighborhood were once rare.

Grow seeds and bulbs on the windowsill and immerse her in the joy of flowers blooming.

Get up early and go for a dawn chorus walk to listen to the voices of birds returning to your neighborhood each spring.

Have a sleepover outside.

Create a habit of collecting moments that make her—and you—feel hopeful.

ACKNOWLEDGMENTS

THIS BOOK WAS written in a few intense months in late 2019/early 2020, but the thoughts that informed it were developed over many years of collaboration. A tremendous collective of individuals has gifted me with kindness, provocative questions, challenging critiques, and opportunities to learn from their expertise. I owe a true debt of gratitude to: the Rachel Carson Center for Environment and Society in Germany; the Rockefeller Foundation Bellagio Center in Italy; Arteles Creative Center in Finland; the Cairns Institute in Australia; the Social Ecology Lab at Stanford University; and Hedgebrook in the USA for welcoming me as a fellow into their vibrant international communities of scholars, policy makers, writers, and other creative thinkers committed to positively impacting the world.

Like good folk who come out in foul weather to line the routes of marathons with encouraging cheers—and loyal running partners who help the miles slip by—many remarkable individuals have created opportunities for me to teach, research, and give keynote speeches; lent me their houses to write in; designed my website; shared podcast chats; and provided strategic support to amplify environmental hope. Thank you to Heather Koldewey, Nancy Knowlton, Elisabeth Whitebread, Cynthia Vernon, Christof Mauch, Rick Kool, Enid Elliot, Kara Shaw, Joy Beauchamp, Nicholas Stanger, Nicole Ardoin, Matthias Neill, Anne Ylvisaker, Paul Fleischman, Carol Diggory Shields, Carrie Parker, Cheryl Joseph, Debbie Basham, Buzz Joseph, Roxane Buck-Ezcurra, Jennifer Good, Karen Kelsey, Alison Kelsey, Janice Harper, Betty Scannell, Ursula Münster, Jim Covel, Ximena Waissbluth, Haley Guest, Eric Higgs, Judy Long, Tom Long, Scout, Simon Chapman, Elizabeth Carrington, Shannon Carnazzo, Kara Lemire, Kathryn Cook, Ken Andrews, Ginny Broadhurst, Serina Allison, Ryan Hilperts, Sarah J. Ray, Jennifer Atkinson, Maria Francesca Troup, Kit Pearson, Katherine Farris, Ellen Field, and Bob Stevenson.

Special thanks to Mary Beth Leatherdale for your friendship, brilliant mind, and thought-filled readings of the draft manuscript. And to Andy Johnson for living this hope journey for so many years and for reviewing many iterations of this book. My son Kip, as luck would have it, is one of the fastest, most voracious readers I know. I cannot thank him enough for his support and for his insightful critiques.

I am indebted to Adrienne Mason and the team at *Hakai Magazine* for granting permission for me to use relevant sections of my *Hakai* articles in *Hope Matters*. I excerpted various portions from "Now, It's Personal" (hakaimagazine.com/features/now-its-personal/, published on April 22, 2015); "The Rise of Ocean Optimism" (hakaimagazine.com/features/rise-ocean-optimism, published on June 8, 2016); "The Power of Compassion" (hakaimagazine.com/features/

power-compassion/, published on August 15, 2017); and "How Animals Code Their Kids for Survival" (hakaimagazine.com/features/how-animals-code-their-kids-for-survival/, published on April 30, 2019).

Over the past decade I have spoken with thousands of students, some as young as preschoolers, others in graduate school, who have bravely shared their feelings and their thoughts about hope and the state of the planet. This book is far richer as a result of those conversations.

Thank you to Rob Sanders for unleashing the creative talent at Greystone Books, and especially to Paula Ayer for her discerning edits and strong commitment to making this book the best it could be.

I know, in my heart, that I will never tire of championing hope. I am more thankful than I can express to my daughter Esmé, who has travelled to conferences, school workshops, and public events all over the world and listened to me talk about hope more than any other person should or could in her young life, and yet still generously helped me with every aspect of creating this book.

A year ago, I took to sleeping outside so I could nod off to the sound of wind and awake to the dawn chorus—and lots of times, rain and rats! The writing of this book was powered by the renewable gifts of sunshine, owl hoots, and the company of a loyal dog.

NOTES

*Unless otherwise noted, all online sources were consulted
between December 1, 2019, and March 1, 2020.*

INTRODUCTION

1. Gillian Flaccus, "West Coast Fishery Rebounds in Rare Conservation 'Home Run,'" *Associated Press News,* December 25, 2019, apnews.
   com/6e7d1ae45aaa7c92a850ae2f70408e51.

2. "Migrating Birds 'Surf the Green Wave,' Tracking Vegetation Across
   Continents," YouTube video, 1:55, posted by American Association
   for the Advancement of Science, January 6, 2017, youtube.com/
   watch?v=lE5V2WMO0DE.

3. Tessa Stuart, "Three Cheers for Birders," *Audubon,* March 13, 2015,
   audubon.org/news/three-cheers-birders.

4. Barbara L. Fredrickson et al., "A Functional Genomic Perspective on
   Human Well-Being," *PNAS* 110, no. 33 (August 13, 2013): 13684–89,
   pnas.org/content/110/33/13684.

CHAPTER 1

1. Aritz Parra and Frank Jordans, "'The Point of No Return Is No Longer
   Over the Horizon': UN Chief Delivers Dire Climate Change Warning,"
   USA Today, December 4, 2019, usatoday.com/story/news/world/2019/
   12/01/un-chief-guterres-climate-change-warning/4347755002/.

2. Johana Kotišová, "The Elephant in the Newsroom: Current Research on Journalism and Emotion," *Sociology Compass* 13, no. 5 (March 2019), doi.org/10.1111/soc4.12677.

3. Laurie Fickman, "Hope Is a Key Factor in Recovering From Anxiety Disorders," University of Houston, October 14, 2019, uh.edu/news-events/stories/2019/october-2019/101419-hope-anxiety-gallagher.php.

4. Madeha Umer and Dely Lazarte Elliot, "Being Hopeful: Exploring the Dynamics of Post-traumatic Growth and Hope in Refugees," *Journal of Refugee Studies*, fez002 (February 2019), doi.org/10.1093/jrs/fez002.

5. Rosie Mestel, "The Imagination Effect: A History of Placebo Power," *Knowable Magazine,* October 25, 2017, knowablemagazine.org/article/mind/2017/imagination-effect-history-placebo-power.

6. Alia Crum, "Harnessing the Power of Placebos," June 7, 2016, TEDMED video, 15:18, tedmed.com/talks/show?id=621415.

7. Alia Crum, "The Science of How Mindset Transforms the Human Experience," YouTube video, 5:08, posted by World Economic Forum, February 21, 2018, youtube.com/watch?v=vtDYtwqKBI8.

8. Alia J. Crum and Ellen J. Langer, "Mind-Set Matters: Exercise and the Placebo Effect," *Psychological Science* 18, no. 2 (2007): 165–71, dash.harvard.edu/handle/1/3196007.

9. Christopher Shea, "Mindful Exercise," *New York Times Magazine,* December 9, 2007, nytimes.com/2007/12/09/magazine/09mindfulexercise.html.

10. Bradley P. Turnwald et al. "Increasing Vegetable Intake by Emphasizing Tasty and Enjoyable Attributes: A Randomized Controlled Multisite Intervention for Taste-Focused Labeling," *Psychological Science* 30, no. 11 (2019): 1603–15, doi.org/10.1177/0956797619872191.

11. Kate Anderton, "Evocative Labels Can Get People to Choose and Consume More Vegetables," *News Medical Life Sciences,* October 3, 2019, news-medical.net/news/20191003/Evocative-labels-can-get-people-to-choose-and-consume-more-vegetables.aspx.

12. Jeremy Howick, "What's the Point in Being Positive?" Jeremy Howick (website), July 11, 2017, jeremyhowick.com/whats-point-positive/.

13. Walter A. Brown, "Expectation, the Placebo Effect and the Response to Treatment," *Rhode Island Medical Journal* 98, no. 5 (May 2015): 19–22, rimed.org/rimedicaljournal/2015/05/2015-05-19-cont-brown.pdf.

14. Joseph Stromberg, "What Is the Nocebo Effect?" *Smithsonian Magazine*, July 23, 2012, smithsonianmag.com/science-nature/ what-is-the-nocebo-effect-5451823/.

15. Alexander Pattillo, "Too Much Bad News Can Make You Sick," *CNN Health*, June 1, 2018, cnn.com/2018/06/01/health/bad-news-bad-health/index.html.

16. Steven Pinker, "The Media Exaggerates Negative News. This Distortion Has Consequences," *The Guardian*, February 17, 2018, theguardian.com/commentisfree/2018/feb/17/steven-pinker-media-negative-news.

17. Jeffrey Gottfried, "Americans' News Fatigue Isn't Going Away— About Two-Thirds Still Feel Worn Out," *Pew Research Center Fact Tank*, February 26, 2020, pewresearch.org/fact-tank/2020/02/26/almost-seven-in-ten-americans-have-news-fatigue-more-among-republicans/.

18. Chloe Reichel, "The Mental Health Effects of Climate Change," *Shorenstein Center on Media, Politics and Public Policy*, September 5, 2019, journalistsresource.org/studies/environment/climate-change/ mental-health-climate-change-research/.

19. Ted MacDonald, "Traditionally Underreported Climate Issues Were Highlighted During the Covering Climate Now Initiative. Broadcast TV News Should Take Note," *Media Matters for America* October 11, 2019, mediamatters.org/diversity-discrimination/climate-issues-were-highlighted-during-covering-climate-now-initiative-and.

20. Elizabeth Arnold, "Doom and Gloom: The Role of the Media in Public Disengagement on Climate Change," *The Shorenstein Center on Media, Politics and Public Policy*, May 29, 2018, shorensteincenter.org/ media-disengagement-climate-change/.

21. "Why 'Solutions Journalism' Matters With David Bornstein," in *Demystifying Media at the University of Oregon*, podcast audio, soundcloud. com/demystifying-media/david-bornstein.

22. Murray A. Rudd, "What a Decade (2006–15) of Journal Abstracts Can Tell Us About Trends in Ocean and Coastal Sustainability Challenges

and Solutions," *Frontiers in Marine Science* 4 (May 31, 2017), frontiersin. org/articles/10.3389/fmars.2017.00170/full.

23. "Annual Global Road Crash Statistics," Association for Safe International Road Travel, asirt.org/safe-travel/road-safety-facts/.

24. David Shepardson, "Major Commercial Airline Deaths Fell by 50% in 2019 Despite a High-Profile Boeing 737 Max Crash," *Business Insider*, January 1, 2020, businessinsider.com/major-commercial-plane-crash-deaths-worldwide-fell-by-more-than-50-in-2019-group-2020-1.

25. Denise Robbins, "This New Study Shows How the Media Makes People Climate Change Cynics—and What They Can Do Differently," *Media Matters in America*, September 18, 2015, mediamatters.org/ new-york-times/new-study-shows-how-media-makes-people-climate-change-cynics-and-what-they-can-do.

26. Arnold, "Doom and Gloom" (see note 20).

27. Anthony O. Scott, "Stuck in Steerage for the Postapocalypse," *New York Times*, June 26, 2014, nytimes.com/2014/06/27/movies/in-snowpiercer-the-train-trip-to-end-all-train-trips.html.

28. Alise Bulfin, "Popular Culture and the 'New Human Condition': Catastrophe Narratives and Climate Change," *Global and Planetary Change* 156 (September 2017): 140–46, sciencedirect.com/science/article/pii/s0921818116303307.

29. Shauna Doll and Tarah Wright, "Climate Change Art: Examining How the Artistic Community Expresses the Climate Crisis," *The International Journal of Social, Political, and Community Agendas in the Arts* 14, no. 2 (2019): 13–29, cgscholar.com/bookstore/works/climate-change-art.

30. Sarah Cascone, "Can Art Change Minds About Climate Change? New Research Says It Can—But Only if It's a Very Specific Kind of Art," *artnet News*, July 26, 2019, news.artnet.com/art-world/art-climate-change-opinions-research-1610659.

31. Jennifer K. O'Leary et al., "The Resilience of Marine Ecosystems to Climatic Disturbances," *BioScience* 67, no. 3 (February 2017): 208–20, academic.oup.com/bioscience/article/67/3/208/2900174.

32. Anthony Leiserowitz et al., "Climate Change in the American Mind: October 2017," *Yale Program on Climate Change Communication*,

November 16, 2017, climatecommunication.yale.edu/publications/
climate-change-american-mind-october-2017/.

CHAPTER 2

1. Yian Yin et al., "Quantifying the Dynamics of Failure Across Science,
    Startups and Security," *Nature* 575 (2019): 190–94, nature.com/articles/
    s41586-019-1725-y.

2. "Does Fear Motivate Workers—Or Make Things Worse?" *Knowl-
    edge@Wharton,* December 4, 2018, knowledge.wharton.upenn.edu/
    article/fear-motivate-workers-make-things-worse.

3. David Wallace-Wells, "The Uninhabitable Earth," *New York Magazine,*
    July 10, 2017, nymag.com/intelligencer/2017/07/climate-change-
    earth-too-hot-for-humans.html.

4. Shoba Sreenivasan and Linda Weinberger, "Fear Appeals," *Psy-
    chology Today,* September 18, 2018, psychologytoday.com/ca/blog/
    emotional-nourishment/201809/fear-appeals.

5. Robert A. C. Ruiter et al., "Sixty Years of Fear Appeal Research: Cur-
    rent State of the Evidence," *International Journal of Psychology* 49, no. 2
    (2014): 63–70, onlinelibrary.wiley.com/doi/full/10.1002/ijop.12042.

6. John Upton, "Media Contributing to 'Hope Gap' on Climate
    Change," Climate Central, March 28, 2015, climatecentral.org/news/
    media-hope-gap-on-climate-change-18822.

7. Andrew Revkin, "Most Americans Now Worry About Climate
    Change—and Want to Fix It," *National Geographic,* January 23, 2019,
    nationalgeographic.com/environment/2019/01/climate-change-
    awareness-polls-show-rising-concern-for-global-warming/.

8. Jacob Poushter and Christine Huang, "Climate Change Still Seen as
    the Top Global Threat, but Cyberattacks a Rising Concern,"
    Pew Research Center, February 10, 2019, pewresearch.org/global/
    2019/02/10/climate-change-still-seen-as-the-top-global-threat-but-
    cyberattacks-a-rising-concern/.

9. Natalie Pierce, "It's Not Just the Protests. Here's How Young Peo-
    ple Are Helping the Planet," *World Economic Forum,* April 18, 2019,
    weforum.org/agenda/2019/04/its-not-just-the-protests-heres-how-
    young-people-are-fighting-for-the-planet/.

10. Emanuela Barbiroglio, "Generation Z Fears Climate Change More Than Anything Else," *Forbes*, December 9, 2019, forbes.com/sites/emanuelabarbiroglio/2019/12/09/generation-z-fears-climate-change-more-than-anything-else/#22f607c4501b.

11. Lindsay P. Galway et al., "Mapping the Solastalgia Literature: A Scoping Review Study," *International Journal of Environmental Research and Public Health* 16, no. 15 (2019), mdpi.com/1660-4601/16/15/2662/.

12. Ted Scheinman, "The Couples Rethinking Kids Because of Climate Change," *BBC*, October 1, 2019, bbc.com/worklife/article/20190920-the-couples-reconsidering-kids-because-of-climate-change.

13. Susie E. L. Burke, Ann V. Sanson, and Judith Van Hoorn, "The Psychological Effects of Climate Change on Children," *Current Psychiatry Reports* 20, no. 35 (April 11, 2018), doi.org/10.1007/s11920-018-0896-9.

14. Jamie Arndt et al., "The Urge to Splurge: A Terror Management Account of Materialism and Consumer Behavior," *Journal of Consumer Psychology* 14, no. 3 (2004): 198–212, sciencedirect.com/science/article/abs/pii/s1057740804701485.

15. Diana Ivanova et al., "Environmental Impact Assessment of Household Consumption," *Journal of Industrial Ecology* 20, no. 3 (2016): 526–36, onlinelibrary.wiley.com/doi/abs/10.1111/jiec.12371.

16. Christine Ro, "The Harm From Worrying About Climate Change," *BBC Future*, October 10, 2019, bbc.com/future/article/20191010-how-to-beat-anxiety-about-climate-change-and-eco-awareness.

17. Anna Zhelnina, "The Apathy Syndrome: How We Are Trained Not to Care About Politics," *Social Problems* (July 3, 2019): 1–21, doi.org/10.1093/socpro/spz019.

18. Esteban Ortiz-Ospina and Max Roser, "Trust," Our World in Data, 2020, ourworldindata.org/trust.

19. Association of Clinical Psychologists UK, "ACP-UK Supports the Call for Clinical Psychologists to Speak Out on Climate Change," ACP-UK, September 2019, acpuk.org.uk/climate_change/.

20. Adam Mayer and E. Keith Smith, "Unstoppable Climate Change? The Influence of Fatalistic Beliefs About Climate Change on Behavioural Change and Willingness to Pay Cross-Nationally," *Climate Policy* 19,

no. 4 (2019): 511–23, tandfonline.com/doi/abs/10.1080/14693062.2018.
1532872.

21. Connie Roser-Renouf et al., "The Genesis of Climate Change Activism: From Key Beliefs to Political Action," *Climatic Change* 125 (2014): 163–78, climatechangecommunication.org/wp-content/uploads/2016/03/May-2014-The-Genesis-of-Climate-Change-Activism.pdf.

22. Robin McKie, "Climate Change Deniers' New Battle Front Attacked," *Guardian*, November 9, 2019, theguardian.com/science/2019/nov/09/doomism-new-tactic-fossil-fuel-lobby.

23. Robin Pomeroy, "At Davos, Trump Urges the World to Ignore the 'Prophets of Doom,'" *World Economic Forum*, January 21, 2020, weforum.org/agenda/2020/01/trump-davos-apocalypse-greta-climate/.

CHAPTER 3

1. Damian Carrington, "Climate-Heating Greenhouse Gases Hit New High, UN Reports," *Guardian*, November 25, 2019, theguardian.com/environment/2019/nov/25/climate-heating-greenhouse-gases-hit-new-high-un-reports.

2. Anna Wlodarczyk et al., "Hope and Anger as Mediators Between Collective Action Frames and Participation in Collective Mobilization: The Case of 15-M," *Journal of Social and Political Psychology* 5, no. 1 (2017), jspp.psychopen.eu/article/view/471/html.

3. L. B. Alloy, L. Y. Abramson, and A. Chiara, "On the Mechanisms by Which Optimism Promotes Positive Mental and Physical Health: A Commentary on Aspinwall and Brunhart," in *The Science of Optimism and Hope: Research Essays in Honor of Martin E. P. Seligman*, ed. J. E. Gillham, Laws of Life Symposia series (Templeton Foundation Press, 2000), 201–12, psycnet.apa.org/record/2001-16094-010.

4. Peter Thompson, "The Frankfurt School, Part 6: Ernst Bloch and the Principle of Hope," *Guardian*, April 29, 2013, theguardian.com/commentisfree/belief/2013/apr/29/frankfurt-school-ernst-bloch-principle-of-hope.

5. Greta Thunberg, "You Can't Just Sit Around Waiting for Hope to Come," YouTube video, 1:18, posted by Fridays for Future, fridaysforfuture.org/greta-speeches#greta_speech_feb21_2019.

6. Shane Lopez, "The Science of Hope: An Interview With Shane Lopez," interview by University of Minnesota, audio, 7:27, takingcharge.csh. umn.edu/science-hope-interview-shane-lopez.

7. Fatemeh Bahmani et al. "The Concepts of Hope and Fear in the Islamic Thought: Implications for Spiritual Health," *Journal of Religious Health* 57, no. 1 (February 2018): 57–71, ncbi.nlm.nih.gov/pubmed/28108912.

8. Erik H. Erikson, *Insight and Responsibility: Lectures on the Ethical Implications of Psychoanalytic Insight* (New York: Norton, 1964).

9. Peter Burke, "Does Hope Have a History?" *Estudos Avançados* 26, no. 75 (2012): 207–18, scielo.br/pdf/ea/v26n75/en_14.pdf.

10. Adam D. I. Kramer, Jamie E. Guillory, and Jeffrey T. Hancock, "Experimental Evidence of Massive-Scale Emotional Contagion Through Social Networks," *PNAS* 111, no. 24 (June 2014): 8788–90, pnas.org/content/111/24/8788.

11. Institute of Physics, "The Most Effective Individual Steps to Tackle Climate Change Aren't Being Discussed," *Phys.org*, July 11, 2017, phys.org/news/2017-07-effective-individual-tackle-climate-discussed.html.

12. C. R. Snyder, "Hope Theory: Rainbows in the Mind," *Psychological Inquiry* 13, no. 4 (2002): 249–75, jstor.org/stable/1448867.

13. Steve Westlake, "Climate Change: Yes, Your Individual Action Does Make a Difference," *The Conversation*, April 11, 2019, theconversation.com/climate-change-yes-your-individual-action-does-make-a-difference-115169.

14. "Carbon Majors," Climate Accountability Institute, October 8, 2019, climateaccountability.org/carbonmajors.html.

15. Joana Setzer and Rebecca Byrnes, "Global Trends in Climate Change Litigation: 2019 Snapshot," Grantham Research Institute on Climate Change and the Environment and Centre for Climate Change Economics and Policy, London School of Economics and Political Science,

July 2019, lse.ac.uk/GranthamInstitute/wp-content/uploads/2019/07/
GRI_Global-trends-in-climate-change-litigation-2019-snapshot-2.pdf.

16. Duane Bidwell and Don Batisky, "Identity and Wisdom as Elements
    of a Spirituality of Hope Among Children With End-Stage Renal
    Disease," *Journal of Childhood and Religion* 2, no. 5 (May 2011): 1–25,
    childhoodandreligion.com/wp-content/uploads/2015/03/Bidwell-
    Batisky-May-2011.pdf.

17. "Conversation With David Orr," *Earth Island Journal*, Winter 2008,
    earthisland.org/journal/index.php/magazine/entry/conversation/.

18. Jennifer, "An Interview With Barbara Kingsolver on Community and
    Hope," Chicago Public Library (blog), September 21, 2016, chipublib.
    org/blogs/post/an-interview-with-barbara-kingsolver-on-
    community-and-hope/.

19. Afua Hirsch, "How to Be an Antiracist by Ibram X Kendi Review—
    A Brilliantly Simple Argument," *Guardian*, October 11, 2019,
    theguardian.com/books/2019/oct/11/how-to-be-an-antiracist-by-
    ibram-x-kendi-review.

20. Noah S. Diffenbaugh and Marshall Burke, "Global Warming Has
    Increased Global Economic Inequality," *PNAS* 116, no. 20 (May 14,
    2019): 9808–13, pnas.org/content/116/20/9808.

21. "Maria Popova: Cartographer of Meaning in a Digital Age," interview
    by Krista Tippett, *On Being*, February 7, 2019, onbeing.org/programs/
    maria-popova-cartographer-of-meaning-in-a-digital-age-feb2019/.

22. Katerina Standish, "Learning How to Hope: A Hope Curriculum,"
    Humanity & Society 43, no. 4 (2019): 484–504, journals.sagepub.com/
    doi/abs/10.1177/0160597618814886.

23. Henry A. Giroux, "When Hope Is Subversive," *Tikkun: A Monthly
    Jewish and Interfaith Critique of Politics, Culture and Society* 19, no. 6
    (2004): 38–39, humanities.mcmaster.ca/~girouxh/online_articles/
    Tikkun%20piece.pdf.

24. Jainish Patel and Prittesh Patel, "Consequences of Repression of
    Emotion: Physical Health, Mental Health and General Well Being,"
    International Journal of Psychotherapy Practice and Research 1, no. 3 (Feb-
    ruary 2019), openaccesspub.org/ijpr/article/999.

25. Maria Ojala, "Young People and Global Climate Change: Emotions, Coping, and Engagement in Everyday Life," in *Geographies of Global Issues: Change and Threat*, ed. N. Ansell, N. Klocker, and T. Skelton, vol. 8 of *Geographies of Children and Young People* (Singapore: Springer, 2016), 329–46, doi.org/10.1007/978-981-4585-54-5.

26. E. Kelsey and C. Armstrong, "Finding Hope in a World of Environmental Catastrophe," in *Learning for Sustainability in Times of Accelerating Change*, ed. A. Wals and P. B. Corcoran (Netherlands: Wageningen Academic Publishers, 2012), 187–200, doi.org/10.1177/0973408213495613.

27. Gabriella Borter, "Rising Seas, Stress Levels Spawn Climate Anxiety Support Groups," *Reuters,* October 23, 2019, reuters.com/article/us-climate-change-eco-anxiety/rising-seas-stress-levels-spawn-climate-anxiety-support-groups-iduskbnix21p2.

28. "Draw Down, Act Up!" Inside the Greenhouse: Re-telling Climate Change Stories, insidethegreenhouse.org/content/draw-down-act.

29. Lisa Kretz, "Emotional Solidarity: Ecological Emotional Outlaws Mourning Environmental Loss and Empowering Positive Change," in *Mourning Nature: Hope at the Heart of Ecological Loss and Grief,* ed. A. Consolo and K. Landman (McGill-Queens University Press: 2017), 258–91, 5b614998-db77-4ba8-83e0-fc20e51959d8.filesusr.com/ugd/e288a7_bb776c970fe04b0e832b09e07ae74ced.pdf.

30. "Teck Withdraws Regulatory Application for Frontier Project," Teck Resources Limited, February 23, 2020, teck.com/news/news-releases/2020/teck-withdraws-regulatory-application-for-frontier-project.

31. Weatherhead Center for International Affairs, "Nonviolence in Mass Uprisings," *Harvard Gazette,* January 27, 2020, news.harvard.edu/gazette/story/2020/01/map-shows-how-nonviolent-uprisings-succeed/.

32. Robin Wright, "The Story of 2019: Protests in Every Corner of the Globe," *New Yorker,* December 30, 2019, newyorker.com/news/our-columnists/the-story-of-2019-protests-in-every-corner-of-the-globe.

33. Erin Duffin, "Proportion of Selected Age Groups of World Population in 2019, by Region," *Statista,* September 20, 2019, statista.com/statistics/265759/world-population-by-age-and-region/.

34. "Climate Emergency Declarations in 1,476 Jurisdictions and Local Governments Cover 820 Million Citizens," *Climate Emergency Declaration*, March 21, 2020, climateemergencydeclaration.org/climate-emergency-declarations-cover-15-million-citizens/.

35. "Dame Cicely Saunders: Her Life and Work," St. Christopher's Hospice, stchristophers.org.uk/about/damecicelysaunders.

36. Peter Yuichi Clark, "Two Ministers, 26 Years: A Reflection on Engaged Hope in Cancer," *Journal of Clinical Oncology* 37, no. 29 (August 2019): 2689–92, ascopubs.org/doi/full/10.1200/JCO.19.00472.

37. Václav Havel, *Disturbing the Peace*, trans. Paul Wilson (New York: Knopf, 1990), 182, accessed through Vaclav Havel Library Foundation website, vhlf.org/havel-quotes/disturbing-the-peace/.

38. Viktor E. Frankl, *Man's Search for Meaning* (Boston: Beacon Press, 2006), 135.

39. "VitalTalk Makes Communication Skills for Serious Illness *Learnable*," VitalTalk, vitaltalk.org.

40. "The Climate Clock: Tracking Global Warming in Real Time," climateclock.net.

41. Roshi Joan Halifax, "Transforming Suffering Today: Imagination and the Bodhisattva Attitude," August 29, 2019, video, 43:01, posted by Upaya Zen Center, facebook.com/watch/?v=340832263465309.

42. Roshi Joan Halifax, "The Strange and Necessary Case for Hope," December 19, 2018, video, 1:28:58, posted by Office for Religious Life, Stanford University, religiouslife.stanford.edu/news/roshi-joan-halifax-delivers-contemplation-design-keynote-strange-and-necessary-case-hope.

43. "Growing Old With HIV," *Out in the Open With Piya Chattopadhyay*, CBC Radio, November 29, 2019, cbc.ca/listen/live-radio/1-131-out-in-the-open/clip/15640701-growing-old-with-hiv.

CHAPTER 4

1. Sarah Kaplan, "Bikini Atoll Islanders Forced Into Exile After Nuclear Tests Now Find New Homes Under Threat From Climate Change," *The Independent*, October 28, 2015, independent.co.uk/environment/

climate-change/bikini-atoll-islanders-forced-into-exile-after-nuclear-tests-now-find-new-homes-under-threat-from-a6712606.html.

2. Sam Scott, "What Bikini Atoll Looks Like Today," *Stanford Magazine,* December 2017, stanfordmag.org/contents/what-bikini-atoll-looks-like-today.

3. "Wolves in Chernobyl Could Spread to Other Areas, Help Support Other Populations," *University of Missouri News Bureau,* July 9, 2018, munewsarchives.missouri.edu/news-releases/2018/0709-wolves-in-chernobyl-could-spread-to-other-areas-help-support-other-populations/.

4. Ker Than, "Gorilla Youngsters Seen Dismantling Poachers' Traps—A First," *National Geographic,* July 18, 2012, nationalgeographic.com/news/2012/7/120719-young-gorillas-juvenile-traps-snares-rwanda-science-fossey/.

5. Max Roser and Mohamed Nagdy, "Optimism and Pessimism," Our World in Data, 2020, ourworldindata.org/optimism-pessimism.

6. John Gramlich, "5 Facts About Crime in the U.S.," *Pew Research Center Fact Tank,* October 17, 2019, pewresearch.org/fact-tank/2019/10/17/facts-about-crime-in-the-u-s/.

7. "The Ignorance Project," Gapminder, gapminder.org/ignorance/.

8. Niall McCarthy, "Pet Euthanasia Has Declined Sharply in the U.S.," *Statistica,* April 11, 2018, statista.com/chart/13493/pet-euthanasia-has-declined-sharply-in-the-us/.

9. Jamie Baxter, "Americans' Pet Spending Reaches Record-Breaking High: $95.7 Billion," American Pet Products Association, February 27, 2020, americanpetproducts.org/press_releasedetail.asp?id=205.

10. "Garden to Table," *National Gardening Association Special Report,* 2014, Garden.org.

11. Sarah Marsh, "Indoor Plant Sales Boom, Reflecting Urbanisation and Design Trends," *Guardian,* August 11, 2019, theguardian.com/lifeandstyle/2019/aug/11/indoor-plant-sales-boom-reflecting-urbanisation-and-design-trends.

12. Alexandre N. Zerbini et al., "Assessing the Recovery of an Antarctic Predator From Historical Exploitation," *Royal Society* 6, no. 10 (2019), doi.org/10.1098/rsos.190368.

13. Drew T. Cronin, "Implementing SMART to Conserve Big Cats Globally," *National Geographic,* February 28, 2018, blog.nationalgeographic. org/2018/02/28/implementing-smart-to-conserve-big-cats-globally/.

14. Chris Brown, "Russia's 'Fairy Tale' Siberian Tigers Beating Long Odds for a Comeback," CBC News, March 21, 2019, cbc.ca/news/world/ siberian-tigers-comeback-1.5041469.

15. Michael Standaert, "In New Park, China Creates a Refuge for the Imperiled Siberian Tiger," *Yale Environment 360,* February 1, 2018, e360.yale.edu/features/china-carves-out-a-park-for-the-imperiled-siberian-tiger.

16. Eugenia Bragina, "Russian Tigers and the Struggle to Maintain Connectivity," *Conservation Corridor,* March 15, 2016, conservationcorridor. org/2016/03/russian-tigers-are-fated-by-lack-of-connectivity/.

17. John Metcalfe, "How Do You Stop Drivers From Colliding With Wild Animals?" *City Lab,* September 27, 2017, citylab.com/environment/ 2017/09/how-do-you-stop-drivers-from-smacking-into-wild-animals/ 541047/.

18. "Roadkill Statistics," *High Country News,* February 7, 2005, hcn.org/ issues/291/15268.

19. Cathy Ellis, "Budget 2019 Sets Aside Money for Wildlife Overpass," *Rocky Mountain Outlook,* November 4, 2019, rmotoday.com/canmore/ budget-2019-sets-aside-money-for-wildlife-overpass-1827435.

20. Bill Graveland, "Success of Wildlife Corridors in Banff National Park Has Advocates Wanting More," *National Post,* December 13, 2019, nationalpost.com/pmn/news-pmn/canada-news-pmn/success-of-wildlife-corridors-in-banff-national-park-has-advocates-wanting-more.

21. Starre Vartan, "How Wildlife Bridges Over Highways Make Animals—and People—Safer," *National Geographic,* April 16, 2019, nationalgeographic.com/animals/2019/04/wildlife-overpasses-underpasses-make-animals-people-safer/.

22. Michael A. Sawaya, Steven T. Kalinowski, and Anthony P. Clevenger, "Genetic Connectivity for Two Bear Species at Wildlife Crossing Structures in Banff National Park," *Proceedings of the Royal Society B: Biological Sciences* 281, no. 1,780 (April 7, 2014), doi.org/10.1098/ rspb.2013.1705.

23. James MacDonald, "Do Wildlife Corridors Work?" *JSTOR Daily,* December 12, 2016, daily.jstor.org/do-wildlife-corridors-work/.

24. Iberá Project, "Giant Anteater," last updated February 2019, proyectoibera.org/en/english/especiesamenazadas_osohormiguero. htm.

25. Jonathan Franklin, "Can the World's Most Ambitious Rewilding Project Restore Patagonia's Beauty?" *Guardian,* May 30, 2018, theguardian.com/environment/2018/may/30/can-the-worlds-largest-rewilding-project-restore-patagonias-beauty.

26. "Greta Thunberg and George Monbiot Make Short Film on the Climate Crisis," YouTube video, 3:40, posted by the *Guardian,* September 19, 2019, youtube.com/watch?v=-QOXUXO2ZEY.

27. Sophie Bertazzo, "What on Earth Are 'Natural Climate Solutions'?" *Conservation International,* November 25, 2019, conservation.org/blog/what-are-natural-climate-solutions.

28. Bronson W. Griscom et al., "Natural Climate Solutions," *PNAS* 114, no. 44 (2017): 11645–50, pnas.org/content/114/44/11645.

29. "Restoration and Current Research," National Park Service, nps.gov/olym/learn/nature/restoration-and-current-research.htm.

30. Lynda V. Mapes, "Elwha: Roaring Back to Life," *Seattle Times,* 2016, projects.seattletimes.com/2016/elwha/.

31. Starre Vartan, "Beavers on the Coast Are Helping Salmon Bounce Back. Here's How," *National Geographic,* August 13, 2019, nationalgeographic.com/animals/2019/08/coastal-beavers-help-salmon-recovery-washington/.

32. Dam Removal Europe, "Dam Removal Boosts Rewilding Efforts in the Ukrainian Danube Delta," Dam Removal Europe, November 25, 2019, damremoval.eu/dam-removal-boosts-rewilding-efforts-in-the-ukrainian-danube-delta/.

33. Clare Wilson, "Harbor Seals Are Breeding in the River Thames and Have Had 138 Pups," *New Scientist,* September 2, 2019, newscientist.com/article/2214891-harbour-seals-are-breeding-in-the-river-thames-and-have-had-138-pups/.

34. Sophie Hardach, "How the River Thames Was Brought Back From the Dead," *BBC*, November 12, 2015, bbc.com/earth/story/20151111-how-the-river-thames-was-brought-back-from-the-dead.

35. Tim Folger, "The Cuyahoga River Caught Fire 50 Years Ago. It Inspired a Movement." *National Geographic*, June 21, 2019, nationalgeographic.com/environment/2019/06/the-cuyahoga-river-caught-fire-it-inspired-a-movement/.

36. "Regional Marine Research Institutions," Monterey Bay National Marine Sanctuary, montereybay.noaa.gov/research/resinstitute.html.

37. Francesco Ferretti et al., "Reconciling Predator Conservation With Public Safety," *Frontiers in Ecology and the Environment* 13, no. 8 (2015): 412–17, doi.org/10.1890/150109.

38. Mark S. Lowry et al., "Abundance, Distribution, and Population Growth of the Northern Elephant Seal (*Mirounga angustirostris*) in the United States From 1991 to 2010," *Aquatic Mammals* 40, no. 1 (February 24, 2014): 20–31, doi.org/10.1578/AM.40.1.2014.20.

39. Jen Sawada, "How Shark Sanctuaries Sparked a Conservation Movement," *Pew Charitable Trusts,* April 1, 2019, pewtrusts.org/en/research-and-analysis/articles/2019/04/01/how-shark-sanctuaries-sparked-a-conservation-movement.

40. Luke O'Neill, "The Public Fear Sharks Less When They Understand Their Behaviour," *University of Sydney,* December 12, 2017, sydney.edu.au/news-opinion/news/2017/12/12/the-public-fear-sharks-less-when-they-understand-their-behaviour.html.

41. Joe Roman et al., "Whales As Marine Ecosystem Engineers," *Frontiers in Ecology and the Environment* 12, no. 7 (September 2014): 377–85, doi.org/10.1890/130220.

42. Bloomberg News, "Saving Whales Would Help the Planet More Than Planting Trees, New Study Says," *Washington Post,* November 20, 2019, washingtonpost.com/lifestyle/kidspost/saving-whales-would-help-the-planet-more-than-planting-trees-new-study-says/2019/11/20/1753404a-0704-11ea-ac12-3325d49eacaa_story.html.

43. Ralph Chami et al., "Nature's Solution to Climate Change," *International Monetary Fund: Finance and Development* 56, no. 4 (December

2019), imf.org/external/pubs/ft/fandd/2019/12/natures-solution-to-climate-change-chami.htm#author.

44. "Fin Whale, Mountain Gorilla Recovering Thanks to Conservation Action—IUCN Red List," International Union for Conservation of Nature, November 14, 2018, iucn.org/news/species/201811/fin-whale-mountain-gorilla-recovering-thanks-conservation-action-iucn-red-list.

45. Katharine Rooney, "Whales Are Vital to Curb Climate Change—This Is the Reason Why," *World Economic Forum*, November 29, 2019, weforum.org/agenda/2019/11/whales-carbon-capture-climate-change/.

46. Matt McGrath, "World's Largest Marine Protected Area Declared in Antarctica," BBC News, October 28, 2016, bbc.com/news/science-environment-37789594.

47. Brian Payton, "Marine Protected Areas: May or May Not Include Actual Protection," *Hakai Magazine*, January 7, 2020, hakaimagazine.com/features/marine-protected-area-may-or-may-not-include-actual-protection/.

48. Enric Sala and Sylvaine Giakoumi, "No-Take Marine Reserves Are the Most Effective Protected Areas in the Ocean," *ICES Journal of Marine Science* 75, no. 3 (May–June 2018): 1166–68, academic.oup.com/icesjms/article/75/3/1166/4098821.

49. "Lunchtime Colloquium: Donald Worster," YouTube video, 41:51, posted by the Rachel Carson Center for Environment and Society, January 17, 2014, youtube.com/watch?v=xmEt9hc8420.

50. "Nature's Dangerous Decline 'Unprecedented'; Species Extinction Rates 'Accelerating,'" Intergovernmental Science-Policy Platform on Biodiversity and Ecosystem Services (IPBES), May 7, 2019, ipbes.net/news/Media-Release-Global-Assessment.

51. "Priority Places for Biodiversity," The Half-Earth Project, half-earthproject.org/maps/.

52. "Business Networks in 24 Countries Pledge Immediate Action to Halt the Deterioration of Nature and the Loss of Biodiversity," World Business Council for Sustainable Development, October 14, 2019,

wbcsd.org/Overview/News-Insights/General/News/Business-
networks-in-24-countries-pledge-immediate-action-to-halt-the-
deterioration-of-Nature-and-the-loss-of-biodiversity.

53. Jeff Tollefson, "'Global Deal for Nature' Fleshed Out With Specific
Conservation Goals," *Nature News,* April 19, 2019, nature.com/articles/
d41586-019-01253-z; Nicole Schwab and Kristin Rechberger, "We
Need to Protect 30% of the Planet by 2030. This Is How We Can
Do It," World Economic Forum, April 22, 2019, weforum.org/
agenda/2019/04/why-protect-30-planet-2030-global-deal-nature-
conservation.

54. Linda Etchart, "The Role of Indigenous Peoples in Combating Climate Change," *Palgrave Communications* 3, no. 17,085 (2017), nature.
com/articles/palcomms201785.

55. Alain Frechette et al., "A Global Baseline of Carbon Storage in
Collective Lands," *Rights and Resources Initiative,* September 2018,
rightsandresources.org/wp-content/uploads/2018/09/A-Global-
Baseline_RRI_Sept-2018.pdf.

56. Richard H. Grove, *Green Imperialism: Colonial Expansion, Tropical
Island Edens and the Origins of Environmentalism, 1660–1860* (Cambridge:
Cambridge University Press, 1995).

57. Joe Curnow, "#Fridaysforfuture: When Youth Push the Environmental Movement Towards Climate Justice," *The Conversation,*
September 15, 2019, theconversation.com/fridaysforfuture-when-
youth-push-the-environmental-movement-towards-climate-justice-
115694.

58. Jimmy Thomson, "Rethinking the Colonial Mentality of Our
National Parks," *The Walrus,* November 26, 2019, thewalrus.ca/
rethinking-the-colonial-mentality-of-our-national-parks/.

59. "Edéhzhíe," Decho First Nations, dehcho.org/resource-management/
edehzhie/.

60. "New Indigenous Protected Areas Will Create Largest Protected Area
on Earth," *Newcastle Herald,* October 28, 2019, newcastleherald.com.
au/story/6462124/new-indigenous-protected-areas-will-create-largest-
protected-area-on-earth/.

61. Andrew J. Trant et al., "Intertidal Resource Use Over Millennia Enhances Forest Productivity," *Nature Communications* 7, no. 12491 (2016), nature.com/articles/ncomms12491.

62. Nicole Smith et al., "3500 Years of Shellfish Mariculture on the Northwest Coast of North America," *PLOS ONE* 14, no. 2 (2019), journals.plos. org/plosone/article?id=10.1371/journal.pone.0211194.

63. Hannah Mowat and Peter Veit, "The IPCC Calls for Securing Community Land Rights to Fight Climate Change," *World Resources Institute,* August 8, 2019, wri.org/blog/2019/08/ipcc-calls-securing-community-land-rights-fight-climate-change.

64. Amber Dance, "Ecuador Grants Rights to Nature," *Nature News Blog,* September 29, 2008, blogs.nature.com/news/2008/09/ecuador_grants_rights_to_natur.html.

65. Gwendolyn J. Gordon, "Environmental Personhood," *Columbia Journal of Environmental Law* 43, no. 1 (November 11, 2019), journals.library. columbia.edu/index.php/cjel/article/view/3742.

CHAPTER 5

1.   "Generation Z Wants More Action for a Sustainable Future, Reveals Global Research From Masdar," *Water and Wastewater Asia,* November 17, 2016, waterwastewaterasia.com/en/news-archive/ generation-z-wants-more-action-for-a-sustainable-future-reveals-global-research-from-masdar/527.

2.   Verena Tiefenbeck et al., "Real-Time Feedback Reduces Energy Consumption Among the Broader Public Without Financial Incentives," *Nature Energy* 4, (2019): 831–32, nature.com/articles/ s41560-019-0480-5.

3.   Rosemary Mena-Werth, "Girl Scouts Can Help Parents Make Energy-Saving Decisions at Home, Research Shows," *Stanford News,* July 11, 2016, news.stanford.edu/2016/07/11/energy-savvy-girl-scouts-can-help-parents-make-energy-saving-decisions-home-research-shows/.

4.   Li Jing, "Pollution Makes Beijing 'Almost Uninhabitable to Human Beings,'" *South China Morning Post,* February 12, 2014, scmp.com/ news/china/article/1426587/pollution-makes-beijing-almost-uninhabitable-human-beings.

5. C. Arden Pope III and Douglas W. Dockery, "Air Pollution and Life Expectancy in China and Beyond," *PNAS* 110, no. 32 (August 6, 2013): 12861–62, pnas.org/content/110/32/12861.

6. J. P., "How China Cut Its Air Pollution," *Economist,* January 25, 2018, economist.com/the-economist-explains/2018/01/25/how-china-cut-its-air-pollution.

7. UN Environment, *A Review of 20 Years' Air Pollution Control in Beijing,* United Nations Environment Programme (Nairobi, Kenya: 2019), wedocs.unep.org/bitstream/handle/20.500.11822/27645/airPolCh_ EN.pdf.

8. Michael Greenstone and Patrick Schwarz, "Is China Winning Its War on Pollution?" Air Quality Life Index, March 2018, aqli.epic.uchicago. edu/wp-content/uploads/2018/08/China-Report.pdf.

9. "Beijing Set to Exit List of World's Top 200 Most-Polluted Cities: Data," *Reuters,* September 11, 2019, reuters.com/article/us-china-pollution-beijing/beijing-set-to-exit-list-of-worlds-top-200-most-polluted-cities-data-iduSKCN1VX05Z.

10. Joseph Poore and Thomas Nemecek, "Reducing Food's Environmental Impacts Through Producers and Consumers," *Science* 360, no. 6,392 (June 1, 2018): 987–92, science.sciencemag.org/content/360/6392/987.

11. Olivia Petter, "The Surprising Reason Why Veganism Is Now Mainstream," *Independent,* April 10, 2018, independent.co.uk/life-style/food-and-drink/veganism-rise-uk-why-instagram-mainstream-plant-based-diet-vegans-popularity-a8296426.html.

12. "Grubhub Launches Annual 'Year in Food' Report Highlighting the Top Trends in 2019," Grubhub, December 4, 2019, media. grubhub.com/media/press-releases/press-release-details/2019/ Grubhub-Launches-Annual-Year-In-Food-Report-Highlighting-The-Top-Trends-In-2019/default.aspx.

13. Monica Watrous, "Trend of the Year: Plant-Based Foods," *Food Business News,* December 27, 2019, foodbusinessnews.net/articles/15105-trend-of-the-year-plant-based-foods.

14. Andreas Vou, "Europe Is Going Veg," *European Data Journalism Network,* March 12, 2019, europeandatajournalism.eu/eng/News/ Data-news/Europe-is-going-veg.

15. Pearly Neo, "Five Top Trends Set to Shape the APAC Food and Beverage Sector in 2020," Food Navigator-Asia, January 6, 2020, foodnavigator-asia.com/Article/2020/01/06/Five-top-trends-set-to-shape-the-APAC-food-and-beverage-sector-in-2020.

16. Tracey Phelps, "Protein: A Chinese Perspective," Plant and Food Research, August 2017, plantandfood.co.nz/file/protein-china-perspective.pdf.

17. "The EAT-Lancet Commission on Food, Planet, Health," EAT-Lancet Commission, January 16, 2019, eatforum.org/eat-lancet-commission/.

18. Janet Ranganathan and Richard Waite, "Sustainable Diets: What You Need to Know in 12 Charts," World Resources Institute, April 20, 2016, wri.org/blog/2016/04/sustainable-diets-what-you-need-know-12-charts.

19. Mark Wilson, "KFC's Beyond Meat Chicken Is a Damn Miracle," Fast Company, January 29, 2020, fastcompany.com/90455889/kfcs-beyond-meat-chicken-is-a-damn-miracle.

20. Thomas Franck, "Alternative Meat to Become $140 Billion Industry in a Decade, Barclays Predicts," CNBC, May 23, 2019, cnbc.com/2019/05/23/alternative-meat-to-become-140-billion-industry-barclays-says.html.

21. Rick Stein, "How the Rise of 'Flexitarians' Is Powering Plant-Based Sales," Food Industry Association, October 22, 2019, fmi.org/blog/view/fmi-blog/2019/10/22/how-the-rise-of-flexitarians-is-powering-plant-based-sales.

22. Alissa Link, "Tech Solutions to Reduce Food Waste: Food and Tech Series," Hunter College New York City Food Policy Center, August 6, 2019, nycfoodpolicy.org/food-and-tech-solutions-to-prevent-reduce-food-waste/.

23. Niko Kommenda, "How Your Flight Emits as Much CO_2 as Many People Do in a Year," Guardian, July 19, 2019, theguardian.com/environment/ng-interactive/2019/jul/19/carbon-calculator-how-taking-one-flight-emits-as-much-as-many-people-do-in-a-year.

24. Jordan Davidson, "JetBlue to Be First Carbon Neutral Airline in the U.S.," EcoWatch, January 10, 2020, ecowatch.com/jetblue-carbon-neutral-airline-2644666201.html.

25. Lucy Handley, "Fast-Fashion Retailers Like Zara and H&M Have a New Threat: The $24 Billion Used Clothes Market," CNBC, March 29, 2019, cnbc.com/2019/03/19/fashion-retailers-under-threat-from-24-billion-second-hand-market.html.

26. *2019 Resale Report*, thredUP, 2019, thredup.com/resale.

27. "How Much Do Our Wardrobes Cost to the Environment?" The World Bank, September 23, 2019, worldbank.org/en/news/feature/2019/09/23/costo-moda-medio-ambiente.

28. "UN Helps Fast Fashion Industry Shift to Low Carbon," *United Nations Climate Change News*, September 6, 2018, unfccc.int/news/un-helps-fashion-industry-shift-to-low-carbon.

29. Thomas Barrett, "35% of Microplastics in Oceans Come From Clothing, Research Reveals," *Environment Journal*, October 5, 2018, environmentjournal.online/articles/35-of-microplastics-in-oceans-come-from-clothing-research-reveals/.

30. Ekaitz Ortega and Ruqayyah Moynihan, "'Köpskam', a New Swedish 'Shame of Buying' Trend, Could Spread to Threaten the World's Fashion Market," *Business Insider*, December 29, 2019, businessinsider.com/swedish-koepskam-shame-of-buying-a-threat-to-fashion-market-2019-9.

31. Gordon T. Kraft-Todd et al., "Credibility-Enhancing Displays Promote the Provision of Non-Normative Public Goods," *Nature* 563 (2018): 245–48, nature.com/articles/s41586-018-0647-4.

32. Erika Engelhaupt, "Does Doom and Gloom Convince Anyone About Climate Change?" *Science News*, July 28, 2017, sciencenews.org/blog/science-the-public/new-york-magazine-climate-change.

CHAPTER 6

1. Adam Vaughan, "Recordings Reveal That Plants Make Ultrasonic Squeals When Stressed," *New Scientist*, December 5, 2019, newscientist.com/article/2226093-recordings-reveal-that-plants-make-ultrasonic-squeals-when-stressed/.

2. Mark J. Costello, Robert M. May, and Nigel E. Stork, "Can We Name Earth's Species Before They Go Extinct?" *Science* 339,

no. 6,118 (January 25, 2013): 413–16, science.sciencemag.org/content/339/6118/413.

3. Indiana University, "Earth May Be Home to One Trillion Species," *Science Daily,* May 2, 2016, sciencedaily.com/releases/2016/05/160502161058.htm.

4. Alison Abbott, "Scientists Bust Myth That Our Bodies Have More Bacteria Than Human Cells," *Nature News,* January 8, 2016, nature.com/news/scientists-bust-myth-that-our-bodies-have-more-bacteria-than-human-cells-1.19136.

5. Adnan I. Qureshi et al., "Cat Ownership and the Risk of Fatal Cardiovascular Diseases. Results From the Second National Health and Nutrition Examination Study Mortality Follow-up Study," *Journal of Vascular and Interventional Neurology* 2, no. 1 (January 2009): 132–35, ncbi.nlm.nih.gov/pmc/articles/PMC3317329/.

6. Miho Nagasawa et al., "Oxytocin-Gaze Positive Loop and the Coevolution of Human-Dog Bonds," *Science* 348, no. 6,232 (April 17, 2015): 333–36, science.sciencemag.org/content/348/6232/333.

7. "Researching Urban Coyote Ecology and Behavior. Sharing Findings So We Can Co-Exist," Urban Coyote Research Project, urbancoyoteresearch.com.

CHAPTER 7

1. Jon Wallace, "Pride Tops Guilt As a Motivator for Environmental Decisions," *Princeton Environmental Institute,* February 14, 2018, environment.princeton.edu/news/pride-tops-guilt-as-a-motivator-for-environmental-decisions/.

2. Paul Clolery, "Volunteer Time Value Hits All-Time High," *Non-Profit Times,* April 12, 2019, thenonprofittimes.com/npt_articles/volunteer-time-value-hits-all-time-high/.

3. Aja Romano, "Hopepunk, the Latest Storytelling Trend, Is All About Weaponized Optimism," *Vox,* December 27, 2018, vox.com/2018/12/27/18137571/what-is-hopepunk-noblebright-grimdark.

4. Mohamad Abrash, "'No Place to Go': Syrian Doctor Says More Than 700,000 Idlib Residents Stranded," interview by Carol Off,

As It Happens, CBC, February 12, 2020, cbc.ca/radio/asithappens/
as-it-happens-wednesday-edition-1.5460922/no-place-to-go-syrian-
doctor-says-more-than-700-000-idlib-residents-stranded-1.5461104.

5. Glenn Althor, James E. M. Watson, and Richard A. Fuller, "Global
Mismatch Between Greenhouse Gas Emissions and the Burden of
Climate Change," *Scientific Reports* 6, no. 20,281 (2016), nature.com/
articles/srep20281.

6. Christina Starmans, Mark Sheskin, and Paul Bloom, "Why People
Prefer Unequal Societies," *Nature Human Behaviour* 1, no. 0082 (April 7,
2017), nature.com/articles/s41562-017-0082.

7. Jamie Margolin, "I'm Not Only Striking for the Climate," *New York
Times*, September 20, 2019, nytimes.com/2019/09/20/opinion/
climate-strike.html.

8. K. N. C, "How to Increase Empathy and Unite Society," *Economist*,
June 7, 2019, economist.com/open-future/2019/06/07/how-to-
increase-empathy-and-unite-society.

9. "Small Scale Interventions Have Become a Popular Way to Enhance
Public Life in Cities. How Do They Influence Our Feelings and
Behaviour?" Happy City, thehappycity.com/project/happy-streets/.

10. Menchi Liu and Emily Valente, "Mindfulness and Climate Change:
How Being Present Can Help Our Future," *Psychology International Newsletter*, October 2018, apa.org/international/pi/2018/10/
mindfulness-climate-change.

11. "The Cooperative Human," *Nature Human Behaviour* 2 (2018): 427–28,
nature.com/articles/s41562-018-0389-1.

12. Mariusz Zieba et al., "Coexistence of Post-traumatic Growth and Post-
traumatic Depreciation in the Aftermath of Trauma: Qualitative and
Quantitative Narrative Analysis," *Frontiers in Psychology* 10 (March 29,
2019), frontiersin.org/articles/10.3389/fpsyg.2019.00687/full.

13. Panu Pihkala, "The Cost of Bearing Witness to the Environmental
Crisis: Vicarious Traumatization and Dealing with Secondary
Traumatic Stress Among Environmental Researchers," *Social Episte-
mology* 34, no. 1 (2020): 86–100, tandfonline.com/doi/abs/10.1080/
02691728.2019.1681560.

14. Lee Rowland and Oliver Scott Curry, "A Range of Kindness Activities Boost Happiness," *The Journal of Social Psychology* 159, no. 3 (2019): 340–43, tandfonline.com/doi/abs/10.1080/00224545.2018.1469461.

15. Lauren Turner, "Why Being Kind Could Help You Live Longer," BBC News, November 11, 2019, bbc.com/news/world-us-canada-50266957.

16. Kristin Neff, "Self-Compassion," self-compassion.org.

17. James R. Doty, "How Compassion Can Save the World," YouTube video, posted by TEDxMenloCollege, November 22, 2019, 17:04, youtube.com/watch?v=GcfE07A7QUY.

18. "Mental Health in the Workplace," World Health Organization, May 2019, who.int/mental_health/in_the_workplace/en/.

19. Anne Kingston, "The World Is Broken—and Human Kindness Is the Only Solution," *Maclean's*, June 19, 2019, macleans.ca/society/the-world-is-broken-and-human-kindness-is-the-only-solution/.

20. "Animal Lovers' Empathy May Be Hardwired in Their DNA," The Roslin Institute, January 3, 2019, ed.ac.uk/roslin/news-events/latest-news/animal-lovers-empathy-may-be-hardwired-in-dna.

CHAPTER 8

1. "Traumatized Elephants," *Harvard Health Publishing*, March 2014, health.harvard.edu/newsletter_article/In_brief_Traumatized_elephants.

2. Andreea Nita et al., "Collaboration Networks in Applied Conservation Projects Across Europe," *PLOS ONE*, October 10, 2016, journals.plos.org/plosone/article?id=10.1371/journal.pone.0164503.

3. Isabel Hilton and Karl Mathiesen, "Why the Guardian Is Publishing Its Elephant Reporting in Chinese," *chinadialogue*, September 8, 2016, chinadialogue.net/blog/9243-Why-the-Guardian-is-publishing-its-elephant-reporting-in-Chinese/en.

4. Severin Hauenstein et al., "African Elephant Poaching Rates Correlate With Local Poverty, National Corruption and Global Ivory Price," *Nature Communications* 10, no. 2,242 (2019), nature.com/articles/s41467-019-09993-2.pdf.

5. Erik Stokstad, "Elephant Poaching Falls Dramatically in Africa," *Science Magazine*, May 28, 2019, sciencemag.org/news/2019/05/ elephant-poaching-falls-dramatically-africa.

6. Huadong Guo, "Big Earth Data: A New Frontier in Earth and Information Sciences," *Big Earth Data* 1, no. 1–2 (2017): 4–20, doi.org/10.1080/ 20964471.2017.1403062.

7. Hossein Hassani, Xu Huang, and Emmanuel Sirimal Silva, "Big Data and Climate Change," *Big Data and Cognitive Computing* 3, no. 12 (2019), researchgate.net/publication/330831437_Big_Data_and_Climate_ Change.

8. "Climate Emergency Declarations in 1,482 Jurisdictions and Local Governments Cover 820 Million Citizens," *Climate Emergency Declaration*, April 2, 2020, climateemergencydeclaration.org/category/news/.

9. "Global Climate Emergency Declarations," Climate Mobilization Project, accessed March 23, 2020, docs.google.com/document/d/ 1vrLtB_ymknee_mkyL5nsh5aaoemNGcsG9l_mec3jrre/edit.

10. "30 of the World's Largest & Most Influential Cities Have Peaked Greenhouse Gas Emissions," C40 Cities, October 8, 2019, c40.org/ press_releases/30-of-the-world-s-largest-most-influential-cities-have-peaked-greenhouse-gas-emissions.

11. Adrian Cho, "Nobel Prize for the Economics of Innovation and Climate Change Stirs Controversy," *Science Magazine*, October 8, 2018, sciencemag.org/news/2018/10/roles-ideas-and-climate-growth-earn-duo-economics-nobel-prize.

12. "President Trump Wants Out—We Are Still In," We Are Still In, wearestillin.com.

13. Global Commission on the Economy and Climate, *Unlocking the Inclusive Growth Story of the 21st Century*, New Climate Economy, 2018, newclimateeconomy.report/2018/.

14. "Marine Plastics," IUCN Issues Brief, May 2018, iucn.org/resources/ issues-briefs/marine-plastics.

15. Carole Excell et al., *Legal Limits on Single-Use Plastics and Microplastics: A Global Review of National Laws and Regulations*, World Resources Institute and United Nations Environment Programme, December 5,

2018, wedocs.unep.org/bitstream/handle/20.500.11822/27113/plastics_limits.pdf.

16. Doris Knoblauch, Linda Mederake, and Ulf Stein, "Developing Countries in the Lead—What Drives the Diffusion of Plastic Bag Policies?" *Sustainability* 10, no. 6 (2018), mdpi.com/2071-1050/10/6/1994/htm.

17. Laura Parker, "Plastic Bag Bans Are Spreading. But Are They Truly Effective?" *National Geographic,* April 17, 2019, nationalgeographic.com/environment/2019/04/plastic-bag-bans-kenya-to-us-reduce-pollution/.

18. Gregory Owen Thomas et al., "The English Plastic Bag Charge Changed Behavior and Increased Support for Other Charges to Reduce Plastic Waste," *Frontiers in Psychology* 10, no. 266, February 26, 2019, ncbi.nlm.nih.gov/pmc/articles/PMC6399129/.

19. "Parliament Seals Ban on Throwaway Plastics by 2021," European Parliament, March 27, 2019, europarl.europa.eu/news/en/press-room/20190321IPR32111/parliament-seals-ban-on-throwaway-plastics-by-2021.

20. Laurent Lebreton et al., "Evidence That the Great Pacific Garbage Patch Is Rapidly Accumulating Plastic," *Scientific Reports* 8, no. 4,666 (2018), doi.org/10.1038/s41598-018-22939-w.

21. Alan Crawford and Hayley Warren, "China Upended the Politics of Plastic and the World Is Still Reeling," *Bloomberg,* January 21, 2020, bloomberg.com/graphics/2020-world-plastic-waste/.

22. "Seeking End to Loss and Waste of Food Along Production Chain," Food and Agriculture Organization of the United Nations, fao.org/in-action/seeking-end-to-loss-and-waste-of-food-along-production-chain/en/.

23. "New Partnership Aims to Drastically Cut Food Loss and Waste in Indonesia," World Business Council for Sustainable Development, August 30, 2018, wbcsd.org/Programs/Food-and-Nature/Food-Land-Use/News/New-Partnership-Aims-to-Drastically-Cut-Food-Loss-and-Waste-in-Indonesia.

24. Nicholas Boring, "France—New Anti-Waste Law Adopted," The Law Library of Congress, March 20, 2020, loc.gov/law/foreign-news/article/france-new-anti-waste-law-adopted/.

25. Douglas Broom, "South Korea Once Recycled 2% of Its Food Waste. Now It Recycles 95%," World Economic Forum, April 12, 2019, weforum.org/agenda/2019/04/south-korea-recycling-food-waste/.

26. Frederica Perera, "Pollution From Fossil-Fuel Combustion Is the Leading Environmental Threat to Global Pediatric Health and Equity: Solutions Exist," *International Journal of Environmental Research and Public Health* 15, no. 1 (2018), mdpi.com/1660-4601/15/1/16/htm.

27. Philip J. Landrigan et al., "The *Lancet* Commission on Pollution and Health," *Lancet*, October 19, 2017, thelancet.com/journals/lancet/article/PIIS0140-6736(17)32345-0/fulltext.

28. Ipek N. Sener, Richard J. Lee, and Zachary Elgart, "Potential Health Implications and Health Cost Reductions of Transit-Induced Physical Activity," *Journal of Transport and Health* 3, no. 2 (June 2016): 133–40, ncbi.nlm.nih.gov/pmc/articles/PMC4917017/.

29. Richard Florida, "The Global Mass Transit Revolution," *City Lab*, September 20, 2018, citylab.com/transportation/2018/09/the-global-mass-transit-revolution/570883/.

30. Prachi Bhardwaj and Shayanne Gal, "The Number of Bike-Sharing Programs Worldwide Has Doubled Since 2014—and the Number of Public Bikes Has Increased Almost 20-Fold," *Business Insider*, July 3, 2018, businessinsider.com/bike-sharing-programs-doubled-since-2014-public-bikes-charts-2018-7.

31. University of Colorado Denver, "Cycling Lanes Reduce Fatalities for All Road Users, Study Shows," *Science Daily*, May 29, 2019, sciencedaily.com/releases/2019/05/190529113036.htm.

32. "Africa and Asia to Lead Urban Population Growth in Next 40 Years—UN Report," *UN News*, April 5, 2012, news.un.org/en/story/2012/04/408132-africa-and-asia-lead-urban-population-growth-next-40-years-un-report.

33. Shally Seth Mohile, "64% of India's GenZ Questions Need to Own a Car: Deloitte Global Survey," *Business Standard*, January 20, 2020, business-standard.com/article/economy-policy/64-of-india-s-genz-questions-need-to-own-a-car-deloitte-global-survey-120011900748_1.html.

34. Åsa Aretun and Susanne Nordbakke, *Developments in Driver's License Holding Among Young People. Potential Explanations, Implications and Trends*, VTI report 824A, 2014, diva-portal.org/smash/get/diva2: 734375/FULLTEXT01.pdf; Peter Barrett, "Teens Putting the Brakes on Driving," *Royal Auto*, June 12, 2019, racv.com.au/royalauto/moving/news-information/teenage-driving-licences.html.

35. Erin Blakemore, "A Car-Free Day in Paris Cut Pollution by 40 Percent," *Smithsonian Magazine*, October 9, 2015, smithsonianmag.com/smart-news/when-paris-banned-cars-day-pollution-dropped-40-percent-180956880/.

36. Adele Peters, "What Happened When Oslo Decided to Make Its Downtown Basically Car-Free?" *Fast Company*, January 24, 2019, fastcompany.com/90294948/what-happened-when-oslo-decided-to-make-its-downtown-basically-car-free.

37. *Sustainable Transport*, Institute for Transportation & Development Policy, December 2018, itdp.org/wp-content/uploads/2019/01/ST_30_FINAL_.pdf.

38. "Madrid Low Emission Zone Reinstated After Protests," BBC News, July 5, 2019, bbc.com/news/world-europe-48886405.

39. "A Portrait of New York City 2018," Measure of America of the Social Science Research Council, June 21, 2018, measureofamerica.org/portrait-nyc/.

40. Theodore A. Endreny, "Strategically Growing the Urban Forest Will Improve Our World," *Nature Communications* 9, no. 1,160 (2018), nature.com/articles/s41467-018-03622-0.

41. Theodore A. Endreny et al., "Implementing and Managing Urban Forests: A Much Needed Conservation Strategy to Increase Ecosystem Services and Urban Wellbeing," *Ecological Modelling* 360 (2017): 328–35, researchgate.net/publication/318779295_Implementing_and_managing_urban_forests_A_much_needed_conservation_strategy_to_increase_ecosystem_services_and_urban_wellbeing.

42. Monica Tan, "More Trees on Your Street Means Fewer Health Problems, Says Study," *Guardian*, July 10, 2015, theguardian.com/society/2015/jul/10/more-trees-on-your-street-means-fewer-health-problems-says-study.

43. Richard Conniff, "Trees Shed Bad Rap As Accessories to Crime," *environment Yale*, 2012, environment.yale.edu/envy/stories/trees-shed-bad-wrap-as-accessories-to-crime.

44. David Christian Rose et al., "Calling for a New Agenda for Conservation Science to Create Evidence-Informed Policy," *Biological Conservation* 238 (October 2019), sciencedirect.com/science/article/pii/s0006320719306111.

45. Thomas Elmqvist et al., "Benefits of Restoring Ecosystem Services in Urban Areas," *Current Opinion in Environmental Sustainability* 14 (June 2015): 101–8, sciencedirect.com/science/article/pii/s1877343515000433.

46. Mathew P. White, "Spending at Least 120 Minutes a Week in Nature Is Associated With Good Health and Wellbeing," *Scientific Reports* 9, no. 7,730 (2019), nature.com/articles/s41598-019-44097-3.

47. Jeffrey Goldstein, James K. Hazy, and Benyamin B. Lichtenstein, "The Innovative Power of Positive Deviance," in *Complexity and the Nexus of Leadership* (Palgrave Macmillan: 2010).

48. Joshua E. Cinner et al., "Bright Spots Among the World's Coral Reefs," *Nature* 535 (2016): 416–19, nature.com/articles/nature18607.

49. Martin Kowarsch and Jason Jabbour, "Solution-Oriented Global Environmental Assessments: Opportunities and Challenges," *Environmental Science and Policy* 77 (November 2017): 187–92, sciencedirect.com/science/article/pii/s1462901117308353.

50. Melynda Fuller, "Gen Z Favors News From Social Media, Digital-First Pubs," *Publishers Daily*, June 26, 2019, mediapost.com/publications/article/337519/gen-z-favors-news-from-social-media-digital-first.html.

51. James Painter et al., *Something Old, Something New: Digital Media and the Coverage of Climate Change*, Reuters Institute, 2016, reutersinstitute.politics.ox.ac.uk/our-research/something-old-something-new-digital-media-and-coverage-climate-change.

52. Kristin Conrad, "Seafood Watch 2020: Making Sustainable Seafood Choices," *Edible Silicon Valley*, January 13, 2020, ediblesiliconvalley.ediblecommunities.com/eat/seafood-watch-2020-making-sustainable-seafood-choices-Bay-Area.

53. "Global Sustainable Seafood Market to Surpass us$18.63 Billion by 2025," Coherent Market Insights, May 24, 2018, globenewswire.com/news-release/2018/05/24/1511642/0/en/Global-Sustainable-Seafood-Market-to-Surpass-us-18-63-Billion-by-2025-Coherent-Market-Insights.html.

INDEX

DAVID SUZUKI INSTITUTE

The David Suzuki Institute is a non-profit organization founded in 2010 to stimulate debate and action on environmental issues. The Institute and the David Suzuki Foundation both work to advance awareness of environmental issues important to all Canadians.

We invite you to support the activities of the Institute. For more information please contact us at:

David Suzuki Institute
219–2211 West 4th Avenue
Vancouver, BC V6K 4S2
info@davidsuzukiinstitute.org
604-742-2899
www.davidsuzukiinstitute.org

Cheques can be made payable to The David Suzuki Institute.